FINAL WARNING

BOOKS BY THOMAS HAUSER

Non-fiction:

MISSING
THE TRIAL OF PATROLMAN THOMAS SHEA
FOR OUR CHILDREN (with Frank Macchiarola)
THE FAMILY LEGAL COMPANION
THE BLACK LIGHTS

Fiction:

ASHWORTH AND PALMER
AGATHA'S FRIENDS
THE BEETHOVEN CONSPIRACY
HANNEMAN'S WAR
THE FANTASY
DEAR HANNAH

FINAL WARNING

The Legacy of Chernobyl

Dr. ROBERT PETER GALE and THOMAS HAUSER

WARNER BOOKS

A Warner Communications Company

Library of Congress Cataloging in Publication Data

Gale, Robert Peter.
  Final warning.
  Includes index.
  1. Chernobyl Nuclear Accident, Chernobyl, Ukraine,
1986. I. Hauser, Thomas. II. Title.
TK1362.S65G35 1988   363.1'9   87-40606
ISBN 0-446-51409-8

Designed by Giorgetta Bell McRee

To Tamar, who is always there when I need her; and Tal, Shir, and Elan, whose generation must complete what we've begun

RPG

And for David Branson:

"My view is, without deviation, without exception, without any ifs, buts, or whereases, that freedom of speech means that you shall not do something to people either for the views they have or the views they express or the views they speak or write. I am for the First Amendment from the first word to the last. I believe it means what it says."

—Hugo Black

TH

We are speaking on this occasion, not as members of this or that nation, continent, or creed, but as human beings, members of the species man, whose continued existence is in doubt. The world is full of conflicts, and overshadowing all conflicts is the titanic struggle between communism and anti-communism. Almost everybody who is politically conscious has strong feelings about these issues. But we want you, if you can, to set aside such feelings and consider yourselves only as members of a biological species which has had a remarkable history, and whose disappearance none of us can desire. We shall try to say no single word which shall appeal to one group rather than to another. All, equally, are in peril, and if the peril is understood there is hope that we may collectively avert it.

—**Bertrand Russell** and
Albert Einstein,

September 1955

FINAL
WARNING

P · A · R · T
ONE

The
Setting

CHAPTER
1

Mitinskoe Cemetery is a sprawling expanse on the perimeter of Moscow. Old women gather daily at its wrought-iron gates to sell flowers in the shadow of a massive concrete wall. The graves are well-kept; the undulating land changes color with the seasons from green to brown. One hundred and fifty yards inside the main entrance, surrounded by birch saplings, two rows of headstones stand guard over newly turned ground. The back row is full, with sixteen white marble headstones, each one bearing a name, date of birth, and date of death. The front row has eleven markers and, by design, room for more.

Orlov, Varsinian. To the Russian people, their names are well-known. Akimov, Parchuk. They and twenty-seven others are the first victims of Chernobyl. Four of their brethren lie elsewhere. Valeri Khodemchuk, a control room operator, died instantly in the explosion and was interred beneath tons of debris. His remains were never found. Three of his countrymen died soon after and were buried in Kiev. Several hundred others, their bodies ravaged by radiation, were airlifted to Moscow for specialized medical care. Some survived; those who did not repose within Mitinskoe's cold gray walls.

Everyone who ever lives will die. All of us know this. We don't like it, and in different ways we imagine ourselves to be immortal; but each of us is dependent for life upon a chemical balance within our bodies which will ultimately fail. We will die, our children will die, and civilization will live on. This is the assumption we make; that in some recognizable form, civilization and certainly our species will survive.

It's an optimistic assumption. Science tells us that the Earth was formed four and a half billion years ago. Life in the form of groupings of chemicals began five hundred million years later. Two billion years after that, green plants began to convert sunlight and chemicals into more complex creations. Collapsing the Earth's history into a twenty-four-hour day, Neanderthal man emerged less than two seconds ago. The Great Pyramids of Egypt were built in the past one-tenth of a second.

Why these numbers? Because in a secluded corner of Mitinskoe Cemetery, there is a sobering reminder that man is in and of nature, not above it; that our species is mortal. "The history of life on earth," Rachel Carson wrote in *Silent Spring*, "has been a history of interaction between living things and their surroundings. To a large extent, the physical form and habits of the earth's animal life have been molded by the environment. Only within the present century has one species—man—acquired significant power to alter the nature of his world."

We, the authors of this book, believe in science; we believe in technology. We also believe that nuclear fission and fusion are the most dangerous processes known to man. They have the capacity to turn the Earth into a desert or make the deserts bloom. It is a power that humanity dare not abuse.

Already, there is concern that science has gone too far; that in splitting the atom, man has crossed a threshold that threatens us all. When environmental systems reach certain extremes, changes occur rapidly, unpredictably, and often without warning. Stresses become self-generating; natural effects appear on a global scale. In past centuries, man has been able either to counter these trends or live with them. Now, the fear exists that something is happening which could destroy us all. This is why Chernobyl—an obscure town in the Soviet Ukraine—has become part of the world's lexicon.

This book is not a primer on the nuclear era. It is first and foremost a personal memoir about Chernobyl, and an attempt to put the world's worst nuclear power accident in perspective. Some people think the issues surrounding nuclear power are so complex that only experts should resolve them. However, we hold a contrary view. All of us live in a world whose future is very much linked to nuclear technology. As citizens in a democracy, it is our obligation to understand what is involved and to act upon these issues without self-deception, exaggeration, or demagoguery. We think it important for people to understand what is known and not known about nuclear matters. Thus, we ask that you read the following pages carefully. In so doing, you'll find that nuclear principles, while not simple, are accessible to any intelligent person who wants to learn. And as our part of the bargain, we will explain these principles as simply and briefly as possible.

It starts with atoms—the building blocks of nature. Atoms are unimaginably small. One hundred million of them placed side by side would form a line one inch long.

Atoms consist primarily of empty space. Their center is composed of protons and neutrons joined together to form a nucleus, which carries most of the atom's weight. If matter were composed of nothing but densely packed nuclei, an object the size of a penny would weigh forty million tons. Yet the nucleus occupies only about one

one-hundred-thousandth of an atom's volume. The rest of the atom consists of space and tiny electrons which orbit the nucleus much as planets orbit the sun.

All matter is composed of atoms, but not all atoms are alike. Their primary difference is in the number of protons and neutrons that comprise each nucleus. For example, a hydrogen atom always has one proton; oxygen eight; uranium ninety-two. Each atom has the same number of orbiting electrons as it has protons. However, atoms of the same element can have different numbers of neutrons. For example, while uranium has ninety-two protons, one type of uranium atom has 143 neutrons while another has 146. These "isotopes" (as different forms of the same element are called) are known as uranium-235 and uranium-238. (92 protons plus 143 neutrons equals 235; 92 protons plus 146 neutrons equals 238.)

Each atom is held together by what physicists call "the strong force"—the strongest force in nature—a force which until recently made it impossible for man to split atoms apart. "It seems probable to me," Sir Isaac Newton wrote in 1704, "that God in the beginning formed matter in solid masses, impenetrable, movable particles so very hard as never to wear or break in pieces, no ordinary power being able to divide what God himself made once in the first creation."

Newton, of course, was not dealing with physics on an atomic level, but for centuries his view prevailed. Nineteenth-century physicists, building on his logic, developed a corollary theory decreeing that mass and energy were separate closed systems; that neither mass nor energy could be created or destroyed. Then, in 1905, Albert Einstein took issue with those before him. Examining the properties of energy, mass, time, and space, he postulated that inert mass contained large amounts of latent energy which could, if fundamental building blocks were broken apart, be released. He further postulated that, during this process, a certain amount of matter would be converted to energy, and that the amount of energy created would equal the amount of mass lost times the speed of light squared. $E = mc^2$. Energy could be created; matter could be destroyed. Still, it was only theory. No one, including Einstein, could implement the process. However, in 1938, physicists determined that when uranium-238 atoms were bombarded by neutrons, their nuclei, if hit, split apart, releasing energy in accordance with Einstein's theory. Moreover, when these nuclei split, their neutrons shot out at high speed toward nearby atoms, causing these neighboring atoms to split. Thus, physicists were able to conclude that if enough uranium atoms were gathered together and properly triggered, there would be a chain reaction—the rapid-fire "fissioning" of one atom after another—which would go on until the supply of uranium was exhausted. During this process, enormous amounts of energy would be released.

On August 2, 1939, fearful that Hitler's Germany had begun a concerted effort to unlock the secrets of the atom, Albert Einstein wrote to President Franklin D. Roosevelt as follows:

Sir:

Some recent work by E. Fermi and L. Szilard, which has been communicated to me in manuscript, leads me to expect that the element uranium may be turned into a new and important source of energy in the immediate future. Certain aspects of the situation seem to call for watchfulness and, if necessary, quick action on the part of the Administration. I believe therefore that it is my duty to bring to your attention the following facts and recommendations:

In the course of the last four months it has been made probable—through the work of Joliot in France as well as Fermi and Szilard in

America—that it may become possible to set up a nuclear chain reaction in a large mass of uranium, by which vast amounts of power and large quantities of new radium-like elements would be generated. Now it appears almost certain that this could be achieved in the immediate future.

This new phenomenon would also lead to the construction of bombs, and it is conceivable —though much less certain—that extremely powerful bombs of a new type may thus be constructed. A single bomb of this type, carried by boat and exploded in a port, might very well destroy the whole port together with some of the surrounding territory.

In view of this situation you may think it desirable to have some permanent contact maintained between the Administration and the group of physicists working on chain reactions in America. One possible way of achieving this might be for you to entrust with this task a person who has your confidence and who could perhaps serve in an unofficial capacity. His task might comprise the following:

a) to approach Government Departments, keep them informed of the further development, and put forward recommendations for Government action, giving particular attention to the problem of securing a supply of uranium ore for the United States;

b) to speed up the experimental work, which is at present being carried on within the limits of the budgets of University laboratories, by providing funds, if such funds be required, through contracts with private persons who are willing to make contributions for this cause,

8

and perhaps also by obtaining the co-operation of industrial laboratories which have the necessary equipment.

I understand that Germany has actually stopped the sale of uranium from the Czechoslovakian mines which she has taken over. That she should have taken such early action might perhaps be understood on the ground that the son of the German Under-Secretary of State is attached to the Kaiser-Wilhelm-Institute in Berlin where some of the American work on uranium is now being repeated.

Yours very truly,
Albert Einstein

Einstein's letter was carried to Washington by the financier Alexander Sachs, who could not get an audience with the President until October 11th. When at last he did, the letter was dismissed as "premature." The following morning, after pleading for a second meeting, Sachs returned to the White House and told Roosevelt the story of an American inventor who wrote to Napoleon, proposing a fleet of ships that could travel without sails in any weather. Napoleon rejected the idea as impractical. The American was Robert Fulton, whose steamship would have enabled France to conquer England. "Mr. President," Sachs concluded, "personally, I think there is no doubt that sub-atomic energy is available all around us, and that one day man will release and control its almost infinite power."

For a moment, Roosevelt sat silent. Then he responded: "What you are after is to see that the Nazis don't blow us up?"

"Precisely," Sachs answered.

Thus began the Manhattan Project, which culminated in August 1945 with the bombing of Hiroshima and Nagasaki. Those who worked on the enterprise believed that, if Germany manufactured a nuclear weapon before the United States, unparalleled disaster would follow. Their task was formidable, demanding virtually everything known of physics and more. Throughout the early 1940s, Roosevelt authorized the secret expenditure of hundreds of millions of dollars to support the program. The project was kept secret from Congress, most members of the President's cabinet, and Vice President Truman, who did not learn of its existence until April 12, 1945, the day Roosevelt died. Three months later, shortly before dawn on July 16, 1945, the first atomic bomb was tested in the New Mexico desert at Alamogordo.

"We were lying there, very tense," recalled Isidor Rabi. "There were just a few streaks of gold in the east; you could see your neighbor very dimly. Suddenly, there was an enormous flash of light, the brightest light that I think anyone has ever seen. It blasted; it pounced; it bored its way right through you. It was a vision which was seen with more than the eye. It was seen to last forever. Finally it was over, diminishing, and we looked toward the place where the bomb had been; there was an enormous ball of fire which grew and grew and rolled as it grew; it went up into the air, in yellow flashes and into scarlet and green. A new thing had been born; a new control; a new understanding which man had acquired over nature. Then, there was a chill, which was not the morning cold; it was a chill that came when one thought as for instance of my wooden house in Cambridge, and my laboratory in New York, and of the millions of people living around there, and this power of nature which we had first understood it to be—well there it was."

The blast vaporized a hundred-foot tower from which the bomb had hung, and dug a crater 1,200 feet in diameter. At the moment of detonation, it produced temperatures of tens of millions of degrees Fahrenheit—a phenomenon matched only by exploding supernovae. Spoke one witness, physicist George Kistiakowsky, "I am sure that at the end of the world, in the last millisecond of the Earth's existence, the last man will see what we have just seen."

Three weeks later, at 2:47 A.M. on August 6, 1945, the Enola Gay lifted off from an American air base on Tinian Island. At three A.M., crew members began final assembly of an object which one of them described as resembling "an elongated trash can with fins." Under cover of night, the plane flew toward the Japanese mainland, accompanied by two observation aircraft which carried scientific instruments and cameras. At 8:15:17 A.M., the plane's bomb-bay doors opened wide and its cargo fell free above Hiroshima. A switch was triggered; then a second, and a third. Forty-three seconds after it had left the Enola Gay, the bomb exploded.

"In the first millisecond after detonation," it was later recorded, "a pinprick of purplish-red light expanded into a glowing fireball a half mile in diameter; the temperature at its core was 50,000,000 degrees Fahrenheit. At ground zero, some two thousand feet directly underneath the explosion, temperatures reached several thousand degrees, melting the surface of granite. After this first blast of light, the fireball suddenly exploded into a mass of swirling flames and purple clouds out of which came a huge column of white smoke which rose to a level of 10,000 feet, flattened outwards to form a mushroom-shaped cloud, then climbed upwards until it reached a height of 45,000–50,000 feet."

$E = mc^2$.

One gram of mass—half the weight of a butterscotch Lifesaver—had been converted to energy to destroy a city and end two hundred thousand lives. Reflecting on the horror, Albert Einstein later declared, "The atomic bomb has altered profoundly the nature of the world as we know it, and the human race consequently finds itself in a new habitat to which it must adopt its thinking."

At present, there are fifty thousand nuclear warheads in the world—on submarines, in planes, on ships, in missile silos. These weapons have an explosive yield of twenty billion tons of TNT—roughly 1,600,000 times the yield of the bomb that destroyed Hiroshima. Thus, the words of Jonathan Schell: "We live in a universe whose fundamental substance contains a supply of energy with which we can extinguish ourselves. . . . All human powers are overmatched by the universal power that was unleashed in the world when the atom was split."

CHAPTER
2

As World War II drew to a close, life on Earth had changed. For the first time, science had made the energy in mass available to man. In 1946, to deal with this phenomenon, Congress created the Atomic Energy Commission and gave it control over all aspects of nuclear development as part of an absolute governmental monopoly over nuclear processes. The AEC, in turn, supervised construction of several nuclear reactors designed to produce fissionable material for bombs. However, these reactors were also capable of producing an extremely useful by-product—electricity.

In December 1953, President Dwight D. Eisenhower made his now-famous "Atoms for Peace" speech, promoting the peaceful use of nuclear energy before the United Nations General Assembly. One year later, Congress amended the Atomic Energy Act to allow for civilian ownership of nuclear material, and empowered the AEC to license private companies to build and operate nuclear power stations. Still, private enterprise moved slowly. Nuclear power involved unknown safety risks, and few if any electrical utilities were willing to expose their shareholders to the literally billions of dollars in damages that could theoretically be awarded in the event of a nuclear disaster. Thus, in 1957, to encourage the development of nuclear energy, Congress passed the Price-Anderson Act, severely limiting the liability of power plant operators in the event of a nuclear accident. Thereafter, nuclear power evolved—slowly at first, then more rapidly.

The first few nuclear power reactors were small units constructed with direct government subsidies. The next dozen were built by General Electric and Westinghouse as "loss leaders" to gain a foothold in the market. By the mid-1960s, however, the utility industry was able to justify the construction of nuclear power plants without subsidies and on purely economic grounds.

In 1974, Congress abolished the Atomic Energy Commission, whose functions were ultimately divided between the Department of Energy and the Nuclear Regulatory Commission. This commission is now responsible for licensing, supervising, and ensuring the safe operation of nuclear power plants in the United States.

There are many different types of nuclear reactors. No two are exactly alike, and the Chernobyl unit which exploded on the morning of April 26, 1986, was vastly different in design from any power-producing unit in the United States and Europe. Still, a general understanding of how nuclear power systems operate is helpful to anyone interested in the issues surrounding nuclear power.

Most electricity generated in the United States begins with water that is boiled and turns to steam. The steam is piped to a turbine, where it pushes against blades, causing the turbine shaft to rotate. The shaft is connected to a generator. When the generator turns, electricity results. Non-nuclear turbine generators operate on steam from boilers which use fossil fuel—most often coal or oil—to heat their water. Nuclear power systems are fueled by uranium. The fissioning of uranium releases vast quantities of heat, causing water to boil and creating steam.

The key component of a nuclear power unit is the reactor core, where fuel is fissioned. Prior to fissioning, the uranium has been enriched and converted to ceramic-coated pellets approximately four-tenths of an inch in diameter and a half-inch long. The pellets have been inserted into hollow tubes made of a corrosion-resistant alloy called zircaloy. Generally, the tubes are about a half-inch in diameter and twelve feet long. Between forty and fifty thousand pellets are joined in this fashion in a typical reactor core. Visually, it helps to imagine the tubes as cigarettes standing upright in a round container.

Inside the reactor, a sustained nuclear reaction occurs. Uranium atoms in the fuel rods split and give off neutrons whose speed is regulated by water or graphite "moderators" to maximize the splitting of more uranium atoms in a chain reaction. As the atoms split, heat is created. The tubes are constantly immersed in water to absorb heat generated by the fission process. In some nuclear power units, water flowing through the reactor core boils and turns to steam, which is piped directly to the turbine. In others, water passing through the core is kept under pressure so it cannot boil. Instead, the heat is transferred to water in a secondary loop which boils and turns to steam.

Even at its safest, nuclear power involves a delicate balance. The chain reaction must be sustained, but it must also be controlled. If too many atoms fission at once, there will be too much heat for the system to handle. Thus, the primary safety concerns in designing and operating a nuclear reactor are to regulate the rate of fission and prevent overheating of the reactor core. This is accomplished through the use of "control rods" and the reactor's "core cooling system."

Control rods are the primary mechanism for regulating the rate of fission. Most often they are long stainless-steel tubes filled with boron carbide powder, although some reactors employ cadmium or graphite—substances equally capable of absorbing neutrons.

A reactor's energy-release rate is controlled by the positioning (in and out) of the control rods. When control rods are inserted between fuel rod assemblies, the rate of fission decreases because neutrons which would normally split other uranium nuclei are absorbed by the rods, which act like a sponge. Conversely, if a reactor's control rods are withdrawn, the rate of fission increases because more neutrons are free to split neighboring uranium atoms. If all of a reactor's control rods are inserted, the fissioning of atoms ends and the reactor shuts down. However, control rods absorb neutrons, not heat. Thus, a second system is required to keep the reactor core from melting in the intense heat of the fission process. This is known as the core cooling system.

Most reactor cooling systems utilize water. As the reactor's uranium fuel fissions, it creates heat which is conducted through the tubes encasing the fuel pellets. These tubes, in turn, are submerged in water that is kept circulating through the reactor core by cooling pumps. It is this water that cools the zircaloy tubes to keep them from melting.

In theory, nuclear power is close to ideal. It's efficient and reasonably priced. In an era when known oil resources are limited, it offers energy without dependence upon foreign powers. However, even the most ardent proponents of nuclear power acknowledge that some risks attend the process.

As nuclear reactors operate, radioactive wastes accumulate inside the fuel rods. These wastes give off decay heat and must be cooled long after the controlled fission process ends. Problematically, there is, at present, no fully accepted method for disposing of them, and they remain highly radioactive for long periods of time.

Beyond the problem of nuclear waste—and in the short run, more serious—is the danger of radiation leaks from a nuclear reactor. A nuclear power plant reactor cannot explode like an atomic bomb. For a variety of reasons, this is impossible. However, midsize reactors contain a radioactive inventory one thousand times that released at Hiroshima, and the release of even a small portion of this material can cause serious human and environmental damage. To combat the danger, most reactors are encased in stainless-steel pressure vessels which sit within reinforced concrete containment structures. However, large steam explosions or external forces such as bombs or hurricanes can under extreme circumstances breach these defenses.

Then, too, there is the ultimate threat of a reactor meltdown. When a reactor is functioning properly, water passes between its fuel rod assemblies to cool the core. If this cooling system fails—either because of a power failure, pump failure, or a broken water main—a series of back-up cooling systems comes into play. However, if all of these systems fail, a meltdown is possible.

Typically, a reactor core is immersed in water. But if the flow of new cooling water is interrupted, water already in the pressure vessel will heat up and boil away, exposing the core. Should this happen, temperatures inside the reactor will begin to rise, and the zircaloy cladding which encases the fuel pellets will melt. Soon, the uranium fuel will melt too, turning the core into a molten radioactive metal mass. Eventually, the core will decompose into a puddle at the bottom of the pressure vessel. Then, reaching 5,000 degrees Fahrenheit, it will melt through the vessel to the floor of the reactor's outer containment structure. These structures are designed to withstand radiation leaks and explosions, not meltdowns. Reacting chemically, molten fuel will eat through the concrete-and-steel floor of the containment structure and continue downward. Once it was thought the fuel would drop through the Earth virtually forever—thus the term "China syndrome." Now, however, it's believed the fuel would interact with elements in the soil beneath the power station and cause steam explosions before coming to rest about twenty yards underground, encased in a glass bubble formed by its own heat acting upon the soil.

Scientists agree that a meltdown would be an event of catastrophic proportions. A 1965 study by the Atomic Energy Commission estimated that a meltdown and breach of containment would kill 27,000 people, seriously injure another 73,000, and cause seventeen billion dollars in property damage. These figures are adjustable up or down depending on weather conditions, the proximity of the reactor to population centers, the reactor's radioactive inventory, and other factors. Adjusted for inflation, the seventeen-billion-dollar figure would rise today to over sixty billion. Virtually all of this damage would be caused by radiation.

What is radiation? And why is it dangerous?

Radiation is the product of an unstable atomic nucleus. Most naturally occurring isotopes (different forms of the same element) are stable. That is, they have no tendency to break apart. But if neutrons are added or taken away from the nucleus of a stable atom, its energy balance becomes uneven and something must be expelled for it to regain balance. For example, plutonium-239 is an artificial element created by adding a neutron to uranium-238. Plutonium-239, like most artificial isotopes, is unstable. In effect, it wants to become something else, and by expelling two protons and two neutrons from its nucleus, it becomes uranium-235. "Radiation" consists of the particles or waves which are emitted by an unstable atom. "Radioactivity" is the spontaneous decay or disintegration of an unstable atom through particle or wave emission.

In and of itself, radiation is quite common. The Earth was born in a radioactive explosion. We live in a radioactive world. Every man and woman is slightly radioactive, since all living tissue contains traces of radioactivity. Still, despite its ubiquity, radiation is dangerous because the particles and waves emitted can alter other atoms integral and important to living organisms. For example, when two protons and two neutrons are emitted together from a nucleus, they are known as an alpha particle. Alpha particles are relatively heavy and large. They interact dramatically with any atom they hit, dislocating electrons and jarring the nucleus. However, alpha particles have relatively little penetrating power and can be stopped by a sheet of paper or a person's skin. Thus, unless an atom emitting alpha particles is ingested or inhaled into the human body, little damage will be done.

Beta particles are single electrons emitted at high speed from unstable atoms. On impact, they cause less damage to the atoms they hit than would an alpha particle. However, beta particles are capable of penetrating both paper and living tissue. Thus, exposure to beta radiation can cause serious skin burns.

Sometimes unstable nuclei emit bursts of energy in the form of waves traveling at the speed of light. These are gamma rays, which have the power to penetrate anything short of thick slabs of concrete or lead. X-rays are similar to gamma rays, except that they are emitted from the electrons of unstable atoms rather than from their nuclei.

Most radiation in the world today comes from cosmic rays, radioactive rocks present in the Earth's crust, and other natural sources. This background radiation has remained relatively constant over the millennia. However, when man began splitting atoms to build bombs and generate electricity, vast new quantities of radioactive material were created. These radioactive by-products have, for the most part, been isolated from the environment by limiting nuclear weapons tests to underground sites and depositing nuclear wastes in temporary storage areas. On occasion, though, radioactive material from these activities has been released into the atmosphere in the form of radioactive dust—fallout—which circles the globe before being brought back to Earth by wind, rain, or the force of gravity. When this occurs, the potential for death and destruction is enormous.

Radiation is silent, invisible, and odorless. It is not apparent to the human senses and, even if it were, man is largely defenseless in its wake.

The effects of radiation on humans are best-known from studies of the victims of Hiroshima and Nagasaki and a handful of laboratory accidents. Many residents of Hiroshima and Nagasaki who appeared unharmed immediately after the blast soon suffered acute radiation sickness and a general breakdown of bodily functions leading to death. Others appeared in good health for years before developing genetic mutations and cancer. Based on these studies and others that followed, scientists agree that radiation kills, and that exposure to any level of radiation poses certain dangers.

Radiation attacks the human body in one of three ways—outside bombardment, ingestion with food or water, or inhalation into the lungs. Damage results when radioactive substances decay and send particles crashing into cells, causing structural changes.

Generally, radiation injuries fall into two categories. The first type occurs when a large dose of radiation has been received and many cells are affected. Under these circumstances, severe tissue damage and radiation sickness are evident within days. Initial symptoms include nausea, vomiting, dizziness, and headaches. These are followed by hemorrhaging, gastrointestinal problems such as diarrhea, and loss of hair. Finally, as millions upon millions of cells die, the body's tissues and organs are destroyed. If the damage is extensive enough, the victim dies. The severity of the illness varies, based on the level of radiation and the effectiveness with which the body's repair mechanisms counter the damage.

The second type of radiation injury is long-term and results from damage to a single cell. Radioactive particles are all around us, and it is inevitable that a random percentage will be ingested or inhaled. Once this occurs, due to their chemical makeup, the particles settle in certain organs. For example, strontium-90 is chemically similar to calcium, and is therefore incorporated into bone. Iodine-131, which is mistaken by the thyroid gland for normal iodine, collects in the thyroid. Cesium-137, which resembles potassium, accumulates in all body cells. In and of themselves, these radioactive atoms are not harmful. However, if one of them lodges, for example, in a cell in the lungs, decays and releases energy, it can damage a nearby cell. That cell may survive and sit dormant for years, but it has been altered.

Thus, you—the reader of this book—could wake up tomorrow, walk out onto the street, and inhale an atom of plutonium released at Chernobyl. That plutonium atom could rest in your lungs until the year 2000 and then emit an alpha particle, damaging the cell in which it resides or a nearby cell. The growth of every cell in the human body is regulated by genes which control when and how the cell divides. If these regulatory genes are damaged, the result may be uncontrolled cell division. In some circumstances, the result of this division is cancer. Cancers caused by radioactive atoms can take decades to develop. One speck of plutonium, one damaged cell, and any one of us could die.

It's scary; a frightening game of numbers and chance. As a consequence of natural background radiation, all of us are struck by approximately fifteen thousand radioactive particles every second. What protects us is that the probability of any given particle decaying while in our body, irreparably damaging a cell, and causing cancer or some other abnormality is very low. But some particles are more dangerous than others, and some that science has created as by-products of nuclear weapons and nuclear power retain their potentially lethal properties for thousands of years.

The length of time a particular substance remains radioactive is measured by its "half-life." This is the substance's rate of decay, the time it takes half of its radioactive matter to convert to a more stable form by emitting waves and particles. For example, iodine-131 has a half-life of eight days. This means that one ounce of iodine-

21

131 will turn into a half-ounce of iodine-131 and a half-ounce of more stable decay products after eight days. After sixteen days, a quarter-ounce of iodine-131 is left; after twenty-four days, an eighth of an ounce. After 160 days (twenty half-lives), less than one-millionth of an ounce of radioactive iodine-131 remains.

However, other radioactive materials are longer-lasting. Strontium-90 has a half-life of twenty-eight years; cesium-137, thirty-three. This means they are potentially dangerous and must be contained for centuries. Twenty-seven different types of radioactive substances are created during the normal fissioning of uranium in a nuclear power plant reactor. Plutonium—a man-made element that didn't exist until uranium was fissioned to make nuclear weapons and nuclear power—has a half-life of 24,000 years.

Plutonium emits alpha particles. Each atom of plutonium, if inhaled, is capable of causing lung cancer. A plutonium atom released into the atmosphere at Chernobyl might have decayed and become harmless the day it was released. But half of the plutonium atoms released at Chernobyl will remain potentially lethal for 24,000 years, a quarter until the year 50,000. No human intervention can hasten the decay process. This is the inherent difference between nuclear energy and other forms of power. When primitive man extinguished his fires, they were out. Modern-day blast furnaces and jet engines can be turned off at will. Nuclear fission goes on and on. Its benefits are obvious. It is also the most dangerous process known to man. This much was recognized by John F. Kennedy, who, in seeking approval for a treaty that would ban nuclear testing in the atmosphere, warned: "The number of children and grandchildren with cancer in their bones, with leukemia in their blood, or with poison in their lungs is not a statistical issue. The loss of even one human life or the malformation of even one baby who may be born long after we are gone should be of concern to us all. . . . We all inhabit this small planet. We all breathe the same air. And we are all mortal."

CHAPTER

3

Three centuries before Christopher Columbus sighted land in the Americas, long before the Renaissance and Reformation changed Europe, five hundred years before Peter the Great extended the Russian Empire, the town of Chernobyl was founded. For eight hundred years, its people lived off the land, harvesting rye and potatoes, breeding cattle and hogs. The soil was sandy, and much of the surrounding area was marshland unsuited for farming. Still, the village lived on. Napoleon and Hitler invaded Russia and failed. Through famine, plague, and bitter-cold winters, the people of Chernobyl held their ground.

Now Chernobyl is virtually deserted, devoid of human life, surrounded by a silent brown forest. It is one of 179 towns and settlements evacuated in the wake of the worst nuclear accident in history. The men, women, and children who lived there have been told that their homes will be uninhabitable for years to come.

Nuclear power is an unforgiving technology; it allows little room for error. Proponents and opponents alike agree that the nuclear industry cannot tolerate the same risks as other industrial endeavors. A single accident can change the world. Thus, nuclear power plants are painstakingly designed, regularly tested, and equipped with numerous back-up safety devices. The men and women responsible for their existence are, for the most part, sincere, dedicated, intelligent individuals, and the safety of the reactor is their primary concern. However, even at the highest levels of technological expertise, failures occur. People make mistakes. Machines break. And over the years, nuclear power plants have endured thousands of malfunctions—some minor, others not so small.

Prior to Chernobyl, the world's most publicized nuclear power plant accident occurred at Metropolitan Edison's Three Mile Island Unit Number 2 near Harrisburg, Pennsylvania. There, at four A.M. on March 28, 1979, the normal flow of cooling water to the reactor core was accidentally cut off and a valve opened, causing water to drain out of the reactor core. As it was supposed to, the reactor automatically shut down, and water from a backup cooling system was pumped in to keep decay heat from melting the core. However, control-room operators were unaware of the open valve and, believing the reactor to be adequately supplied with cooling water, they shut off the back-up cooling pumps. Two hours passed before the error was discovered. By then, water in the core had begun to boil and the core was partially uncovered. This caused the reactor's uranium fuel rods to overheat, swell, and finally rupture. Some radioactive material was emitted into the containment structure and thousands of gallons of radioactive water were pumped to an auxiliary building which housed water-storage tanks. A small amount of radioactive gas escaped through a charcoal filter into the open air. The average exposure to members of the public was minimal. More disturbing was the fact that a Nuclear Regulatory Commission study later revealed that the reactor had come within sixty minutes of a meltdown.

Four years prior to the accident at Three Mile Island,

another meltdown had threatened at a nuclear power plant in Decatur, Alabama. On March 22, 1975, a twenty-year-old electrician's aide was checking for air leaks in the cable-spreader room at the Brown's Ferry Unit Number 1, operated by the Tennessee Valley Authority. His check consisted of holding a lit candle in various places and watching to see if the flame flickered. During this process, polyurethane insulation around one of the cables caught fire. The resulting blaze raged through the cable-spreader room, where electrical cables from all over the plant converged, and burned out of control for seven and a half hours. Sixteen hundred cables, including six hundred integral to plant safety systems, were damaged. Virtually all of the safety systems designed to channel cooling water to the reactor core were rendered inoperative.

Every nuclear power plant system carries the potential for failure. Malfunctions can result from errors in design, manufacture, installation, maintenance, or day-to-day operation. Sabotage, earthquakes, tornados, and other natural phenomena carry equal potential for disaster. Brown's Ferry and Three Mile Island were two reminders. Comparable incidents at Detroit Edison's Fermi Unit Number 1 and various nuclear power plants around the world underscore safety concerns. Still, no one died at Three Mile Island. No radiation was released into the atmosphere at Brown's Ferry. And as the years passed, many observers began to question whether the "dangers" of nuclear power were real or imagined. Then came Chernobyl.

During World War II, the United States and Germany were not alone in considering the potential of atomic weapons. As early as 1939, Russian physicists had recognized the possibility of harnessing a nuclear chain reaction. Three years later, Joseph Stalin ordered that high priority be given to investigating the military applications of nuclear fission, and in 1949 the Soviets successfully detonated an atomic explosion. Nuclear parity between the superpowers followed, and thereafter the Soviet government moved to implement an energy policy based in significant part on nuclear fission. By 1984, 10 percent of all electricity generated in the Soviet Union came from nuclear sources, and officials were targeting a 50 percent figure for the year 2000.

The pride of the Soviet nuclear program was a planned six-unit complex at Chernobyl. Chernobyl Unit Number 1 went into service in 1977, and was followed by Units 2, 3, and 4 in 1978, 1981, and 1983 respectively. Units 5 and 6 were scheduled to come on line in 1988. The Chernobyl reactors were similar to one another in design, but differed from units in the United States in two significant respects. First, rather than using water as a moderator, the Chernobyl units employed graphite. And as a consequence of this and several other design features, chain reactions within the reactor were more likely to run out of control in the event of a loss-of-coolant accident. And second, while the Chernobyl units were protected by strong walls, they were housed in buildings which lacked reinforced concrete domes typical of Western containment structures. Whether either of these factors contributed to the scope of the disaster which ultimately occurred is subject to dispute, given the magnitude of the Chernobyl explosion.

The chain of events leading to disaster began on the morning of April 25, 1986. Unit Number 4 was scheduled to be taken out of service for routine maintenance, and the plant's electrical engineers wanted to conduct a test to determine how long the turbine-generators would continue to produce electricity to run the water pumps necessary to cool the reactor after the normal electrical supply had been interrupted.

At one A.M., the reactor's power level was lowered to prepare for the test. Then, over the next twenty-four hours, technicians systematically disconnected power regulation and emergency cooling systems which would have automatically shut the reactor down and interfered with the test. Finally, at 1:23 on the morning of April 26, 1986—twenty-four hours after preparation for the test had begun—the flow of steam to the turbine was halted. Almost immediately, the cooling pumps slowed, diminishing the flow of cooling water to the reactor core. Normally, at this point, the reactor would have shut down, but the automatic shutdown system was one of six safety mechanisms that had been deliberately disconnected. Within seconds, there was a massive heat buildup in the reactor core, triggering an uncontrolled chain reaction. Power surged. At 1:23 A.M., the Unit Number 4 reactor exploded.

There were two blasts, three seconds apart. The first was caused by steam, the second by steam or hydrogen which had formed when the fuel-rod cladding began to melt and interacted with water in the pressure vessel. The reactor core was torn apart. Its thousand-ton coverplate was propelled upward, causing the building roof above the reactor to collapse. A deadly plume of radioactive material—more than was released at Hiroshima and Nagasaki—shot into the air, forming a fiery image above the roof before dispersing into the atmosphere. Exposed to intense heat and open air, the graphite moderator began to burn. Radioactive water gushed into the reactor hall. Hot chunks of fuel and metal landed on what was left of the building roof and the roofs of adjacent buildings. Thirty fires began to burn.

Within minutes, the nuclear plant's firefighting unit was on the scene—twenty-eight men under the command of Major Leonid Telyatnikov. "You had the impression you could see the radiation," Telyatnikov said later. "There were flashes of light springing from place to place, substances glowing, luminescent, a bit like sparklers."

Telyatnikov ordered a "stage three" alarm, the highest for Soviet firefighters, summoning 250 reserves from as far away as Kiev. Then he and the few men available began working desperately to halt the fires. Their primary concern was that Unit Number 4 shared a ventilation

27

system with, and was housed in the same reactor hall as, Unit Number 3. If the fire spread and Unit 3 went up in flames, the disaster would double in magnitude.

A dozen firemen in the Unit 3 reactor block attacked the blaze with hand-held extinguishers. As that struggle progressed, Telyatnikov led six men up a hundred-foot ladder to the collapsed roof of the reactor hall. Because of the heat, what was left of the asphalt roof had begun to melt. With each step, the firemen's boots sunk into the bitumen, and they had to strain to pull free. Poisonous fumes made breathing difficult; visibility was near zero. Water that was poured on the flames turned instantly to scalding radioactive steam.

Firefighting units from nearby towns began to arrive at three-thirty A.M. Still, the fires raged. "We knew about the radiation," Telyatnikov said later. "We were trying to get the fire before the radiation got us. We are firemen. This is what we were trained for. We are supposed to fight fires. We knew we must stay to the end. That was our duty."

Shortly before dawn, all fires except one in the Unit 4 reactor core (which would burn for weeks) had been extinguished. The reactor hall was in shambles. Of the seven men who fought the blaze on the building roof, all but Telyatnikov (who was hospitalized for three months) would die as a result of radiation exposure.

The disaster at Chernobyl set in motion an extraordinary chain of events with profound social, political, and scientific consequences. A personal memoir of these events follows.

PART
TWO

Chernobyl
Memoir

CHAPTER

I'm a doctor. That means I should be able to deliver a baby, suture cuts, take out an appendix, and perform all the other basic skills of a physician. I'm also an internist, and beyond that, a hematologist, an oncologist, and an immunologist. Carrying specialization further, most of my professional life has been devoted to seeking a cure for leukemia.

Leukemia is a cancer of the bone marrow, which results in the production of excess numbers of white blood cells. Bone marrow is spongy tissue located primarily in the sternum, ribs, backbone, and pelvis. Each day, a healthy adult's bone marrow produces 210 billion red blood cells, which carry oxygen to the body's tissues; 105 billion white ones, which fight infection; and 175 billion platelets, which prevent excessive bleeding and allow the blood to clot. Leukemia destroys this balance. When a person has leukemia, his or her bone marrow produces large numbers of abnormal white cells. Instead of the body maintaining a healthy balance, it's flooded with white cells, which upset the balanced production of blood. The end result, absent successful medical treatment, is that leukemia victims die.

Dealing with leukemia patients has given me a special perspective on life. I've learned what's important. Watching other children die, I consider it a blessing that my own children are healthy. I'm less inclined to be upset if my car doesn't start. I realize it doesn't matter whether or not there's orange juice in the refrigerator when I wake up in the morning.

Most of my work—laboratory research, teaching, and patient care—is conducted at the UCLA Medical Center. Each day I deal with people who are suffering from leukemia. They've come to UCLA to participate in extraordinarily complex, life-threatening medical measures in the hope that these procedures will lead to what most of us consider a natural right—a long life.

I'm absolutely convinced that no one should die of leukemia. And I refuse to concede a patient's life without doing something if the patient wants to be saved. Sometimes my thinking runs against the grain. Often, the easiest thing in medicine is to do nothing; to say this patient has a fatal disease and is going to die. That way, the family accepts death, and if by chance the patient lives, everyone is happy. The problem is, too often negative thinking becomes a self-fulfilling prophesy. If the patient expects death, the nurses expect death. Soon, the doctors expect and accept death too. I'd rather fight gravity and try to change the course of nature, so a twenty-year-old can reach age seventy-five.

In 1945, the year I was born, virtually everyone with leukemia died within two or three years. Today, two-thirds of the children and one-third of the adults who develop it are cured. My professional goal is to reduce leukemia to the status of a non-fatal disease. On April 25, 1986, I was working toward that end. Much of my afternoon was spent discussing a research project with a biochemist from the Johnson & Johnson Biotechnology Center. At six P.M., I telephoned my wife, Tamar, to say

I'd be home from the hospital in ten minutes. At seven o'clock, following the usual unexpected delays, I called Tamar a second time to say I was really and truly ready to leave for home. By then, our daughters, Tal and Shir, and our two-year-old son, Elan, were pleading famine, and Tamar was threatening to serve dinner without me because "the children cannot wait any longer." I promised to leave the hospital immediately, at which point a research assistant with a predictably urgent problem appeared at my door.

Finally, I made it home. During dinner, the children took turns reporting on their day. Afterward, Tal and Shir finished their homework while Elan and I played with his fleet of plastic trucks, which outnumbered those employed by the Allies in support of the invasion at Normandy. Once the kids were in bed, I spent several hours reading medical journals, then went to sleep at midnight. I was unaware that, halfway around the world, during the preceding hours, an event had occurred which would change my life.

THREE DAYS LATER

Usually, I wake up without an alarm clock at six A.M. The house is quiet; the children are still asleep. I'll shave, take a swim in the backyard pool, do a series of exercises, and place telephone calls to Europe and the eastern United States, where the workday has begun. I don't read a daily newspaper. I suppose, in some ways, that's limiting, but I'm in Europe and Asia on business at least a dozen times each year. I deal one on one with scientists and medical experts from all over the world and, as a result, I think I have as much if not more insight into global affairs than most people. Also, I listen to the news while shaving each morning.

On the morning of April 29, 1986, I learned about Chernobyl. The radio report was fragmentary. There had been a serious accident at a Soviet nuclear power plant north of Kiev. Officials in West Germany and Sweden reported that Soviet authorities had asked for advice on how to extinguish a radioactive fire, and speculation was rife that a nuclear reactor was burning out of control. Unconfirmed reports placed the death toll at near two thousand.

Listening to the news, I finished shaving and considered the implications of what I'd just heard. Radiation most seriously affects cells that divide quickly—those making up bone marrow, hair follicles, and the gastrointestinal tract. Given what was known about the accident at Chernobyl, it seemed likely there would be many radiation victims whose bone marrow had been destroyed. And without functioning bone marrow, within weeks a person will die from bleeding or infection similar to that of a person with untreated leukemia.

All of this was familiar terrain to me because of my work with leukemia patients. Many children and adults with leukemia are cured by chemotherapy. For those who aren't, bone marrow transplantation is a source of hope. Essentially, it's a procedure by which someone suffering from leukemia is given a new blood-producing system. First, the patient's own bone marrow, which is manufacturing diseased blood cells, is destroyed. This is accomplished through chemotherapy followed by a high dose of radiation strong enough to kill the remaining bone marrow cells without excessively damaging the body's other vital organs. Next, a quart of bone marrow (10 percent of the body's total) is extracted by syringe from a bone marrow donor. Once withdrawn, the bone marrow is strained to remove clumps of fat and pieces of bone, and infused through a vein into the leukemia patient's bloodstream much like an ordinary blood transfusion. If all goes well, the transplanted bone marrow cells are carried by the blood to cavities inside the recipient's bones, where they begin to manufacture healthy new blood cells.

Ultimately, success is dependent upon two critical factors. First, the donor's bone marrow must be genetically matched with that of the recipient. If it isn't, the transplant may fail to engraft or may engraft transiently and then be rejected. Either of these events is usually fatal. Or alternatively, white blood cells in the graft may attack the recipient in what is the opposite of graft rejection. This complication is known as "graft-versus-host disease," and is fatal in approximately half of the cases in which it occurs. And second, a successful transplant requires extraordinarily sophisticated post-operative care in the weeks following transplantation, when the new bone marrow has yet to produce enough cells to reconstitute the body's defenses against infection. During this period, the patient is highly susceptible to infection, hemorrhaging, and other complications.

The dilemma is obvious. Some people who have transplants die earlier than they would without them; but the rest are cured. All I can do when patients with leukemia come to me is give them the statistics and let them decide. I tell them as simply and plainly as possible: This type of treatment offers a 50 percent chance of cure. This one is 30 percent. If you definitely want to be here a year from now, don't have a bone marrow transplant. But if you're willing to risk not being here in a few months for the chance to live a long healthy life, you should consider it.

Every person who undergoes a bone marrow transplant is fighting to stay alive. However, leukemia patients can opt or not opt for a transplant as a matter of choice. Pondering the implications of Chernobyl, I realized that many Soviet citizens had no option. Their bodies had already been radiated; their bone marrow had already been destroyed. Without new bone marrow, they would die. However, in the same breath, I wondered whether the Soviets—or any people—were capable of responding to a disaster of this magnitude. Bone marrow transplants are incredibly complex. The transplant itself is only part of the problem. The intensive medical care required after the transplant is an even more limiting factor. In the United States, virtually every bone-marrow-transplant patient bed is currently occupied. Worldwide, only two thousand transplants are performed each year.

I knew generally what the Soviets had been doing in the field of bone marrow transplantation. They had two, possibly three, transplant units, and performed only a handful of bone marrow transplants annually. If the accident was anything close to what was being reported, they'd need help; any country would. At the very least, it seemed to me that there should be an offer of assistance. Why? Because innocent people were dying. It was like any national disaster, except this disaster happened to be in my area of expertise. It was an opportunity to help.

One of the organizations I'm proudest of being associated with is the International Bone Marrow Transplant Registry. Established in 1971, the registry receives reports from approximately 150 transplant centers in sixty countries around the world. Detailed information on every bone marrow transplant performed at participating units is forwarded to the registry, computerized, and disseminated to improve future operations. Dr. Mortimer Bortin is the registry's director. I am chairman of the advisory committee.

Minutes after learning of the Chernobyl disaster, I telephoned Bortin at his office in Milwaukee. Although the Soviet Union did not participate in the registry (they have since joined), Bortin agreed that help should be offered. He, in turn, telephoned the other members of the advisory committee to determine what level of resources were available, how many patients they could take, and, most important, whether or not there was a consensus to help. By nine A.M., I had Bortin's authorization on behalf of the registry to offer assistance to the Soviets.

The next question was how to extend the offer. By now, there were reports on the radio that the State Department had offered humanitarian and technical aid to the Soviets and that the offer had been declined. I considered seeking out the National Cancer Institute or the National Institute of Health as conduits, since they fund much of our work at the registry and UCLA. However, both of these organizations are government bodies and were unlikely to be accepted by the Soviets. Simply sending my own telex on behalf of the registry to the Soviet Embassy in Washington would be futile. I had to consider what would be going on in the Soviet Union in the midst of this crisis. And with that in mind, I turned to one of the more remarkable men I've known—Dr. Armand Hammer.

Hammer was born in New York to Russian immigrant parents on May 21, 1898. After graduating from medical school at Columbia University, he journeyed to the Soviet Union in 1921 to help combat a typhus epidemic, and met briefly with Lenin, who suggested that the Soviets needed American business contacts, not doctors. Thereafter, Hammer arranged to export American wheat to the famine-plagued Ural Mountains in exchange for minerals and furs, which he sold abroad. From 1921 through 1930, he lived in the Soviet Union as a capitalist businessman, representing thirty-eight United States companies in commercial dealings with the Soviets. He also developed an asbestos mine in the Ural Mountains and opened a pencil factory in Moscow. In 1930, he sold these holdings to the Soviet government and bought a host of czarist art treasures, which he marketed through the newly created Hammer Galleries in New York.

In the decades that followed, Hammer prospered in the United States in fields as diverse as distilling, banking, cattle breeding, and broadcasting. In 1956, he purchased a substantial interest in a failing oil company—Occidental Petroleum—for $100,000 and turned it into the eighth-largest oil company in the United States. Then in 1961, after an absence of three decades, he returned to Moscow on a trade mission for John F. Kennedy. Since then, he has maintained cordial relations with every Soviet head of state and, in the mid-1970s, negotiated a twenty-year twenty-billion-dollar fertilizer trade agreement on behalf of Occidental with the Soviet government.

Hammer is a complex individual. Like many successful businessmen, he understands power and uses it often, sometimes harshly. Some people feel his actions are motivated by flagrant self-promotion, and he's made numerous enemies on the road to success, although seldom has he made an enemy of those in power. Still, I'm convinced that in his dealings with the Soviets, Hammer has been motivated by more than a desire for profit and power. He genuinely wants peace between the world's superpowers, and has devoted a substantial portion of his life to that end. In 1986, at age eighty-eight, he flew 250,000 miles and made seven trips to the Soviet Union. As Walter Cronkite once observed, he is "an almost unique bridge between communism and capitalism."

I met Hammer for the first time by chance in 1978 while attending a conference on genetics at Moscow University. It was my first visit to the Soviet Union and, before returning to the United States, I decided to travel to Odessa and Leningrad. When I reached Odessa, the hotel was in turmoil. Some famous American had arrived with a huge entourage and displaced several dozen guests, although my own room reservation was intact. That evening, I went down to the dining room, and there was Armand Hammer, who had just completed negotiations to build a fertilizer plant on the outskirts of Odessa. I introduced myself as a fellow Californian; he was polite; and that was that.

Then, in 1984, we met more formally. Over the years, Hammer has been active in seeking a cure for cancer. He's donated millions of dollars to research and served since 1981 as chairman of President Reagan's Cancer Advisory Panel. One afternoon in September of 1984, I received a telephone call from Paul Terasaki, the director of UCLA's tissue-typing laboratory. Hammer was considering a request for financial support from a scientist who was studying bone marrow transplantation at the Hadassah Hospital in Jerusalem, and had asked Terasaki to review the application. Paul didn't feel he had the expertise to review it properly and asked if I'd go with him to meet with Hammer.

We met at Occidental Petroleum's headquarters in Westwood. I hadn't had a chance to study the proposal in detail, but gave Hammer some initial thoughts and he responded, "Look, we're going to Jerusalem this Saturday morning. Why don't you come with us?"

The idea was appealing. I'd spent a year in Israel doing research at the Weizmann Institute, and my wife, Tamar, is Israeli. But this was Thursday evening, and my passport was at the Chinese Consulate in San Francisco, since I'd just applied for a visa to visit China.

"Don't worry about a passport," Hammer told me. "My driver will pick you up on Saturday morning." And sure enough, when Hammer's driver came to my home on Saturday, he had an emergency passport issued by the State Department. That impressed me. I can't imagine trying to accomplish something like that on my own in one working day and succeeding.

The trip itself was vintage Hammer. We flew on Oxy One—the flagship aircraft in Occidental Petroleum's fleet of private jets—a prerogative he's accorded as Chairman of the Board and Chief Executive Officer. En route to Jerusalem, we stopped in Vienna so Hammer could meet with former Austrian Chancellor Bruno Kreisky. Then we went to London. Hammer had purchased a letter that Lenin wrote requesting membership in the Royal Academy of Sciences, and wanted to collect it in person to give to the Soviets as a gift. At one point, we picked up a group of Arab businessmen who had moved from Beirut to Athens and were involved in construction in Brunei. One of their plans was to build an oil pipeline through the Negev Desert along the border between Israel and Jordan. Both countries, they reasoned, would have a vested interest in maintaining the pipeline's security, since they'd want to keep the oil flowing. These same businessmen also had plans to build a major medical center on the West Bank of the Jordan River to upgrade health care for Arabs, which they felt would be a stabilizing influence. When we got to Israel, in addition to pledging $250,000 to Hadassah Hospital's bone-marrow-transplant research program, Hammer met for an hour with Israeli Prime Minister Shimon Peres and more extensively with former Prime Minister Menachem Begin.

During the course of the trip, I got to know Hammer reasonably well. Subsequently, I traveled with him to meetings of the President's Cancer Advisory Panel and developed increasing respect for his intellect. I also knew that, except for Presidents and Secretaries of State, he was probably the best-known American in the Soviet Union, a multimillionaire who mingled with top members of the Soviet hierarchy, the Soviet Union's "favorite capitalist." So at nine-thirty A.M. on April 29, 1986, I telephoned his office at Occidental Petroleum to seek his help in sending a message to Moscow.

As it turned out, Hammer wasn't in Los Angeles. He was in Washington, D.C., readying to attend the National Gallery opening of an exhibition of paintings from the Pushkin and Hermitage, two of the foremost museums in the Soviet Union. Hammer had arranged for the Soviets to send on tour forty-one Impressionist works, thirty-three of which had never been shown in the United States. In return, the Soviets were to receive forty Impressionist and Post-Impressionist paintings from the National Gallery and Hammer's private collection, "Five Centuries of Masterpieces."

I reached Hammer by telephone at the Madison Hotel and explained that international resources would probably be necessary to respond to an accident of the magnitude of Chernobyl. He listened and asked several questions. Only once did he disagree. That was when I suggested that one option might be taking patients out of the Soviet Union to hospitals in Europe. "Forget that," he told me. "The Soviets will never allow it." At the end of our conversation, he concurred that the idea of medical assistance was worth pursuing, but indicated he didn't want to do anything that ran counter to our own government's policies. Thus, later that day, he broached the subject at a meeting with four members of the Senate Foreign Relations Committee—Chairman Richard Lugar, Albert Gore, Ted Stevens, and Clairborne Pell. Each senator was supportive in response. Only then did Hammer send the following letter to Mikhail Gorbachev:

Dear Mr. General Secretary:

I was saddened to hear of the accident at the Chernobyl nuclear power station near Kiev. We were particularly concerned to hear that there may have been exposure of some of the population to nuclear material.

As you know, one of the possible consequences of this type of exposure can be irreversible damage to the blood and bone marrow which can be fatal. Previous nuclear reactor accidents have been associated with this type of serious problem which results in death of victims some two weeks after exposure with no immediate symptoms.

Individuals exposed to bone-marrow-lethal doses of nuclear radiation can be potentially rescued by the transplantation of bone marrow cells from a suitable donor. Donors might be relatives or volunteers chosen from a computer-base donor pool.

Dr. Robert Peter Gale, Chairman of the International Bone Marrow Transplantation Registry, Professor of Medicine and Chairman of the Bone Marrow Transplantation Unit at the University of California at Los Angeles (UCLA), has offered to mobilize the resources of the United States and international transplantation centers to provide possible assistance in identifying individuals who might require bone marrow transplants as a life-saving procedure. I am well acquainted with Dr. Gale on both a personal and professional basis and know of his world renowned work in my capacity as President Reagan's adviser on cancer matters. He and his team are also prepared to help identify potential bone marrow donors either from within the USSR or using both the US and European computer-based donor banks which are currently in existence.

Dr. Gale is prepared to come immediately to the Soviet Union to meet with Soviet nuclear scientists and hematologists to assess the situation and decide on the optimal course of action with the hope of saving the lives of those at risk. I will bear all costs for his efforts, which can be so important to saving the lives of those citizens who have been exposed.

Mr. General Secretary, please accept my profoundest sympathy for this tragedy and my sincerest offer to assist in any way. I am now in Washington and can be reached through the Embassy here.

With warmest regards,

Respectfully,
Armand Hammer

The letter was delivered to Oleg Sokolov (acting Soviet Ambassador to Washington) and telexed to Anatoly Dubrynin (a key foreign policy adviser to Gorbachev and Secretary of the Communist Party's Central Committee in Moscow). Both men were asked to relay the message to Mr. Gorbachev as soon as possible. Meanwhile, the following morning, Senators Lugar, Gore, Stevens, and Pell sent a joint letter to the State Department requesting approval of the medical rescue effort, which was quickly granted. Twenty-four hours later, at six-thirty A.M. on May 1st, I received a telephone call from Rick Jacobs, a special assistant to Dr. Hammer, who reported that the Soviet Embassy had just asked for my home telephone number. At seven o'clock, I received a call from Ambassador Sokolov, who said he had a cable from Moscow asking me to come to the Soviet Union immediately.

"Moscow or Kiev?" I asked.

"Moscow."

It was a brief conversation. Sokolov was very calm and businesslike. Then I hung up, and all hell broke loose.

During the preceding day, I'd mapped out contingency plans for a rescue effort. Any disaster taxes medical resources. Medical systems aren't designed to deal with mass accidents. If a bus turned over in front of the UCLA Medical Center and seventy people were seriously injured, it would be an extraordinarily demanding job simply to stop the bleeding, find seventy beds, and put intravenous lines into each patient. I could appreciate the magnitude of what lay ahead in the Soviet Union.

In the absence of hard data, I decided to consider a range of contingencies and scenarios. If a hundred people needed bone marrow transplants, what would we do? If there were a thousand victims, how could they be treated? What if the area of destruction is this large? What if it's larger? What facilities will be available? If there are transplants, who will the donors be?

Now that the Soviets had approved my coming to Moscow, it was necessary to contact some of the people I might want to have join me. Paul Terasaki was one of the world's foremost experts in tissue typing. In the event that it proved necessary to type hundreds of people preparatory to transplants, he was the obvious first choice. Yair Reisner was a biochemist at the Weizmann Institute in Israel who'd developed a technique for removing cells from bone marrow that might otherwise cause "graft-versus-host disease" in transplant recipients. Assuming the Soviets didn't have perfectly matched donors, he too would be essential. Dick Champlin was an assistant professor of medicine at UCLA and someone I worked with regularly. If I needed hands-on assistance in patient evaluation and care, he was the one.

By mid-morning, I'd contacted Champlin and Terasaki at UCLA. Shortly before noon, I tracked Reisner down in New York. A week before, he'd attended a medical conference in Colorado and, on the way back to Israel, had stopped in Manhattan to visit his sister and some colleagues at the Sloan-Kettering Institute. Basically, my message was the same to each of them. I didn't know what the situation was in Moscow. I didn't know what, if anything, would be required. I didn't know if they'd be asked to come or not. But they should assume they would be, and I wanted them to assemble everything they might need to perform their specific tasks. Terasaki should gather enough tissue-typing material to type two hundred individuals, a number I chose arbitrarily. Champlin should collect all the practical things necessary to perform transplants—needles, anticoagulants, glass beakers. Reisner should prepare for his laboratory work as though there were nothing but an electrical socket in Moscow.

Next, I telephoned Mortimer Bortin at the International Bone Marrow Transplant Registry and alerted him to the need to map out certain data. Following that, I turned my attention to the problem of compatible donors. The ideal bone marrow donor is an identical twin, the automatically perfect genetic match. Other siblings offer a 25 percent chance of compatibility. The probability of a genetic match between two unrelated people is about one in ten thousand. One of my fears was that if the Chernobyl disaster was of sufficient magnitude, entire families might have been radiated, ruling out the possibility of using relatives as donors. This, in turn, would require a global search for genetically compatible donors.

Worldwide, approximately 150,000 people have volunteered to serve and have been genetically typed as bone marrow donors. The largest register of these volunteers is maintained by the Anthony Nolan Trust, named after a young English child who died of leukemia for lack of a donor. The Nolan Trust, which was established by funds from his parents, advertises actively and has fifty thousand volunteers whose tissue types are on a computer file. No organization that large exists in the United States, but most major medical centers here have lists of highly motivated people—often policemen or firefighters—who have been tissue-typed for platelet or white blood cell donations and are often willing to serve as bone marrow donors. With time running out, I telephoned John Goldman at the Royal Post Graduate Medical School in London and John Hansen at the University of Washington Medical School in Seattle and asked them to contact all relevant bases in the event that unrelated bone marrow donors were needed. Then I began preparing to go to Moscow.

The first step was to pack. I travel light with a standard package. Blue jeans, jogging clothes, casual shirts, a shaving kit, blue blazer, gray slacks, several white shirts, and a tie. I also packed a small tape deck with speakers and, in the rush of the moment, hit the Beethoven section in our living room very hard.

Somehow during this period, the press had found out I was going to Moscow; I don't know how. The result was that reporters had begun to congregate outside our house, and the telephone line was clogged by incoming calls. Avoiding the media as best I could, I dropped by Tal and Shir's school to kiss them good-bye. I told them that there had been an accident in the Soviet Union, that I was going to help take care of some people who had been injured, and that I'd be home as soon as possible, probably in several weeks. Then I drove to my office at UCLA to pick up some recent publications on radiation and bone marrow transplantation, things the Soviets might not have seen.

Someone from Occidental Petroleum delivered a first-class airplane ticket for Moscow to our house. Just before leaving, I telephoned Hammer. We talked at length, and he said it was crucial for me to gain the Soviets' trust. That meant they had to understand that I was there to help them. He also suggested that, in the beginning at least, I not speak to the press. And he closed by saying that he wanted to be informed of all developments, day or night.

Tamar drove me to the airport, which was swamped with reporters. All I told them was, "I'm going to the Soviet Union on behalf of the International Bone Marrow Transplant Registry. I'll be dealing with Soviet patients who have Soviet doctors. I don't know the magnitude of what I'll find. I'll be there to help as best I can."

In truth, I didn't know much more than that. I knew there had been an accident near Kiev, but I was going to Moscow. I knew the Soviets had asked me to come, but who had asked was still unclear. I wasn't even sure whether they wanted me for medical reasons or simply because an American doctor might somehow bolster their credibility.

At three-thirty P.M., Lufthansa flight 451 took off from Los Angeles, en route to Frankfurt and Moscow. Soon, my initial questions would be answered.

CHAPTER
5

I arrived in Moscow at six P.M. the following night, got off the plane, and walked toward the airport terminal. Just before the gate, two men stood waiting. One of them was tall and heavyset, about forty-five years old, dressed in a distinctly American gray pinstriped suit. The other was about my height, five-foot-nine, with dark hair, wearing gray slacks and a blue blazer. The taller of the two stepped forward to greet me.

"Dr. Gale, I'm Victor Voskresenskiy from the Ministry of Health. It's good to meet you."

Victor, a psychiatrist by training, had left the profession because, as he later told me, "I wasn't doing anything; psychiatry doesn't help anybody." Thereafter, he'd become part of the Soviet political-medical establishment, moving to the United States with his wife and thirteen-year-old daughter to serve as a medical attaché at the Soviet Embassy in Washington. At the time of the Chernobyl accident, he'd been vacationing in the Crimea and was assigned to serve as my guide and interpreter.

The man with Victor was Nikolai Fetisov, also from

the Ministry of Health. In a few minutes, we'd passed through the VIP lounge and were in a black limousine, heading toward the center of Moscow.

The air was cool; it was growing dark. During the ride, I asked if we were going to the hospital, and they said no, I must be tired, and they would take me to my hotel to rest. I told them I'd slept on the plane and was eager to begin work, but Victor explained that a Soviet physician who'd been treating the victims would be at the hotel to brief me.

After a drive of about thirty minutes, we arrived at the Sovietskaya, a diplomatic hotel reserved for special visitors. It was an elegant structure, with elaborately painted ceilings, chandeliers, antique furniture, and original art in the lobby and corridors. Once a private mansion, it had a decidedly regal look, with red carpets and a long white marble staircase leading to the mezzanine level.

In the lobby, I was introduced to Dr. Alexander Baranov, a tall balding man in his late forties with a prominent nose and large ears. Unlike Victor and Fetisov, Baranov wore a Soviet-made suit and glasses. His face was warm with an aura of kindness and caring. He looked like a nice man, and in the months that followed that perception proved accurate.

After I checked in at the main desk, we went upstairs. The Soviets had given me a suite, Number 28, with an anteroom, sitting room, bedroom, and two baths. The furniture was rosewood, with Persian rugs on the floor. There was also a television, radio, and phonograph, but no records. Later I would receive several records as a gift and discover that the phonograph didn't have a needle.

Baranov spoke little English. "Very good to see you; happy to have you here," was about all he said. We settled in the sitting room and then, with Victor translating, Baranov began to describe the situation.

There were, he told me, about thirty seriously injured patients who had been transferred from Chernobyl to a hospital in Moscow. Some of them were suffering from bone marrow failure. Others had serious radiation burns and related problems. The Soviets had performed several bone marrow transplants and were arranging for more potential donors. I offered again to go to the hospital immediately, but Baranov said no, I'd see the patients in the morning. Throughout the conversation, which lasted about an hour, he impressed me as knowledgeable, thoughtful, and very much in control of the situation. He also seemed genuinely happy to see me and wasn't at all defensive about my presence, which I took as a sign of respect not for me personally so much as for American medicine.

Meanwhile, Fetisov had ordered dinner sent up to the room, and a waiter arrived with tongue, cucumbers, tomatoes, mineral water, and brown bread. Baranov said he'd meet me at eight in the morning at the hotel for breakfast; Victor showed me how to use the telephone. Then they left and I was alone. All I'd been told, really, was that thirty patients were in the hospital. Their injuries were serious, and work toward a series of bone marrow transplants had started. With that in mind, I discarded my scenario for dealing with two hundred victims. Even if the numbers I'd been given were inaccurate, it seemed logical to focus on fewer patients.

At ten P.M., I telephoned Hammer. Because of the time differential, it was noon in Los Angeles. I filled him in on what had happened and promised to keep him apprised of future developments. Then I telephoned Tamar and, after that, went to sleep.

The next morning, Saturday, May 3rd, I woke up at six o'clock and went jogging. The truth is, I don't like running. That surprises people, because I jog regularly and have completed several marathons with a best time of three hours and twenty minutes. But I run for my health, not for pleasure.

Outside, on the streets of Moscow, I was the only jogger. And since I was wearing a singlet with USA emblazoned across the front, I felt rather conspicuous. Usually, when I travel, I have a good sense of direction. Here, though, I was a bit disoriented. I wanted to run to Red Square, but after four miles realized that I was going in the wrong direction, toward the airport. At one point, I passed a huge building that was part of the Moscow Petroleum Institute. In conjunction with their May Day celebration, the Soviets had put pieces of colored paper in all the windows on one side of the building, creating a giant revolutionary-style portrait of Lenin. I jog with an FM-radio headset, but on this particular run it didn't work. It seemed unlikely that the KGB was following me, jamming my radio, so I assumed that FM bands were on different frequencies in the Soviet Union. I ran eight or nine miles at a seven-minute-per-mile pace. And as I ran, reflecting on finding myself so suddenly behind the "Iron Curtain," a black spiritual kept running through my mind:

Freedom, freedom, oh freedom over me
And before I be a slave, I be buried in my grave
And go home to my Lord and be free.

Back at the hotel, I showered, dressed, and waited for Baranov. As planned, we met at eight. There were two restaurants in the Sovietskaya, one on the ground floor and one on the mezzanine level. We sat in the downstairs restaurant and waited for fifteen minutes for someone to come to take our order. Finally, Baranov went to look for a waiter, came back, and reported, "Slight technical problem." Apparently, the downstairs restaurant was only open to the general public. Hotel guests had to eat upstairs, although presumably the food came from the same kitchen.

We relocated and were served breakfast rolls with coffee. Baranov spoke as best he could in English, although we were limited in our understanding of each other. Essentially, we were taking the next tentative step in building a friendship, and again he said in a very sweet way, "Very happy to have you here."

Victor joined us after breakfast, and outside the hotel

introduced me to a short stocky man in his late fifties whose name was Surik. For as long as I stayed in Moscow, Surik would be our driver.

The ride to the hospital took about thirty minutes. On the way, I asked if it had a name, and was told, "Yes, it is Hospital Number Six." When we arrived, the outside of the building reminded me of an American public hospital from the 1950s. It was ordinary-looking, made of brown brick, nine stories high, surrounded by a lawn and cast-iron fence. However, that was where the resemblance to hospitals I'd seen ended. The grounds were an armed camp. A hundred soldiers were on the lawn, in and around tents, milling about in various stages of activity. They'd been recruited for security purposes and to serve as hospital orderlies the same way the National Guard might be called out in the event of a comparable disaster in the United States.

For security reasons, the Ministry of Health was required to call ahead and report the license-plate number of each car before its occupants were allowed to enter the hospital. Driving to a rear gate, we encountered a stout elderly woman who refused to let us pass. Flustered, Victor ordered Surik back to the front gate, where we were blocked by another "babushka." Finally, after heated argument highlighted by much shouting on Victor's part, the second babushka relented, and we made our way through the gate, out of the car, and into the hospital. Just beyond the front door, we came to a group of technicians with Geiger counters who measured the level of radioactivity on our shoes and trousers. Many of the Chernobyl patients had been contaminated with radioactive dust when they arrived at the hospital. Others had radioactive blood and urine, and when samples were taken, the radioactivity spread to laboratory equipment as part of an endless chain. Thus, every time we entered or left the hospital, we were monitored for radiation.

Baranov's office was on the sixth floor, and we walked up, since the elevator was slow in coming. I sensed an urgency in the people around us, beyond what was normal for a hospital setting. No one was racing through the corridors with bags of blood and whistles blowing, but clearly this wasn't business as usual.

Baranov was a senior physician in the Department of Hematology. The department chairperson was a doctor named Angelina Guskova. Soon after we reached Baranov's office, she appeared and was introduced. Guskova was short, stocky, sixty years old, and very Siberian-looking with a round wrinkled face and piercing eyes. Her father, grandfather, and great-grandfather had all been doctors. Her primary training was as a neurologist and radiologist, but obviously she knew a lot about hematology. She was also prominent within Soviet medical-political circles, having authored a book, *Workers in the Nuclear Industry*, and several articles on the dangers of nuclear war. Unlike Baranov, who had never been out of the Soviet Union, Guskova had been in the international arena for some time.

With Victor translating, Guskova welcomed me to the hospital. Basically, what she said was, "We're happy to have you. Dr. Baranov will take you to see some of the patients, and we will meet again later on." The conversation lasted several minutes, no more. Most likely, she simply wanted to check me out and let me know who was in charge.

Then Baranov and Victor took me down to the sterile block on the fifth floor. Four patients were ensconced in "life islands" breathing filtered air—four plastic bubbles with one patient in each. Outside the block, there was a red line on the floor and, before crossing over, we put on special gear. First, there were cloth booties, tied at the top of the calf. In the United States, we use disposable booties, but these were reusable. Next, a nurse gave us caps, taking them with forceps from a sterile tin. Finally, we put on surgical masks and gowns.

Inside the sterile block, Baranov introduced me to two more physicians: Giorgi, a large man with a red beard, and Svetlana, an attractive young woman just below Giorgi in the hospital hierarchy. Then we looked at the first patient.

The unit was set up in such a way that patients could be at least partially examined from outside their bubbles without our going in. We could see through the plastic, and the bubbles had portholes with fitted gloves which enabled us to touch and manipulate patients without inducing any germs. The first patient was a fireman, a young man with red spots on his chest and arms. He was lying on a bed, wearing pajama slacks with an open top. Baranov explained that the red spots were burns from beta radiation.

During the next hour, we examined all four patients, two from outside their bubbles and two from within. Prior to entering each bubble, we sterilized our hands with a contraption I'd never seen before. It was the size of a sink, filled with water. Following Baranov's lead, I put my hands in, hit a button with my knee, and an ultrasound field sent a tingling sensation through my hands as it sterilized them. Then a nurse gave me a sterile towel, and the job was done.

Baranov introduced me to each patient in Russian. Clearly, he was moved by their plight and cared deeply for all of them. At the time of the accident, there had been 176 staff members on duty at Chernobyl, with 268 more workers on a night shift at a nearby construction site. Still, the foremost candidates for injury had been firefighters, and the patients I examined my first day were all firemen.

They were surprised and, I think, happy to see me. People who are sick are pleased if an expert is brought in to treat them. I was introduced as "a specialist from the United States," and American medicine is highly regarded in the Soviet Union.

The most visually dramatic injuries I saw were burns. These men had been fighting a fire that reached 5,000 degrees Fahrenheit, with water that became scalding hot and radioactive. They had thermal burns and burns from radiation. In some instances, they'd suffered lung damage from inhaling hot toxic fumes, and each of them had absorbed life-threatening amounts of radiation.

Two of the areas most affected by radiation are the scrotum and the space between the buttocks, because they're damp with high bacterial content. Areas where skin meets skin and glands are present, such as the armpits, are also particularly sensitive. With Baranov doing most of the probing, we examined each patient. I doubt they understood the gravity of their illness. When a person is well and gets sick instantaneously, it takes time to acknowledge and comprehend. And here, there was a particularly strong element of denial because the victims were firefighters. Several days before, they'd been strong, healthy individuals. Now they looked ill, but not critically so to the naked eye. They felt weak, but it was too early for them to have lost any hair. They didn't understand that in all likelihood they would get sicker and weaker and soon die.

After our tour of the sterile block, we returned to the sixth floor, where Guskova and a half-dozen other physicians conducted a review of patient condition and care. The room, which would serve as command headquarters throughout my stay, was long and narrow, wood-paneled, with a large window at one end. It had cabinets and a sink, but the most prominent furnishing was a glass-topped desk with three telephones on top. One was red, one green, and one white. Each seemed to serve a different purpose and, since they were right next to each other, when they rang it was difficult to determine which was which.

The Soviets, I learned, had already performed three bone marrow transplants. In addition, they had sought to save three other victims of the disaster through experimental treatment involving fetal liver cells. In a fetus, during the second trimester, the bones are solid and blood cells are produced in the liver. Thus, for transplant purposes, a fetal liver equals bone marrow. Also, the fetus is immunologically naive, so transplanting fetal liver cells avoids the danger of graft-versus-host disease in the recipient. Several Chernobyl victims had been so seriously exposed to radiation that it was impossible to determine their tissue type and find a matched donor. And for those individuals, the Soviets had proceeded with fetal liver transplants.

Guskova conducted the entire review meeting, sometimes in gruff fashion, always professionally, and with no tolerance for wasted breath. "She is like a Siberian bear," one of her colleagues said later. "Dry, direct, upright, and not very talkative." Baranov, a heavy smoker, chain-smoked throughout—strong-looking Russian cigarettes, equivalent to old Camels. At one point, a nurse came in with coffee, tea, jam, and bread.

What impressed me most about the session was a system the Soviets had developed for determining the amount of radiation absorbed by each patient. This was crucial because with less than a certain radiation dose, a bone marrow transplant is unnecessary. And over a certain dose, it's useless because the patient will die anyway for reasons other than bone marrow failure.

Under Guskova's system, three primary variables were considered: how quickly the victim's white blood cell count dropped; the length of time between exposure and the onset of nausea or vomiting; and the extent of chromosome abnormalities in the patient's blood and bone marrow cells. It was far more sophisticated than any system for radiation scoring that I knew of, and as the weeks progressed it became clear that this hospital was not simply a hematology-oncology unit. It was devoted, among other things, to dealing with radiation accidents. When Guskova said, "This is a beta burn," she knew what she was talking about.

Radiation burns are uncommon. Usually, when one occurs, it's a burn to the hands suffered by someone who has managed to outwit the safety device on a machine used for industrial sterilization purposes. Beta burns are virtually unknown in the United States; I'd never seen one in my life. But Guskova obviously had seen lots of them, and I wondered how she and her colleagues had amassed so much data prior to Chernobyl.

When the meeting ended, Surik drove Victor and me back to the hotel for lunch. Then we returned to the hospital, and Baranov invited me to join him in the operating room to extract bone marrow from a donor.

The surgical suite was on the third floor, and similar to those in the United States. So was the pre-surgical ritual. First we removed all our jewelry, which in my case consisted of a watch that I looped around the string on my surgical pants. Then Baranov and I scrubbed in adjoining sinks, and I carefully followed his style of scrubbing. Initially, we took sponges that had been soaked in an iodine solution and washed our hands, starting with the fingers and working toward the elbows, never going back. The idea is to push germs away from your fingers, only go in one direction. All the nurses were watching me carefully, and I scrubbed for the same length of time as Baranov. Once I'd finished, a nurse laid a towel on my hands and I dried myself the same way I'd scrubbed, working toward the elbows. The nurses gowned us; we put on masks and were gloved.

The donor was a young woman, a sister of the transplant recipient. She had already been anesthetized and was lying on the operating table. Baranov scrubbed her buttocks with a sterile solution, and we made two small incisions with a scalpel, one on each side of her buttocks, to allow access for the needles we'd use in extracting bone marrow. The needles are sealed; otherwise they'd fill up with bone upon insertion. Once inserted into the bone, however, the center seal is removed and a syringe is attached to extract the bone marrow cells. Then both the needle and syringe are withdrawn and handed to a nurse who ejects the contents (about one teaspoon of bone marrow and peripheral blood) into a glass beaker. Every surgeon has a slightly different rhythm, but after a while the nurse you're working with adapts so she's handing you the next needle the moment you're ready.

The process takes about ninety minutes. By the time Baranov and I had finished, each of us had made a hundred insertions. Together, we'd drawn a quart of bone marrow. Then the cells were counted, and the bone marrow was transfused to the recipient by another physician. A second transplant was scheduled for later in the afternoon, but Baranov insisted I'd done enough work, and we went back upstairs to the sixth floor.

It had been a remarkable day. I didn't know all of what was going on, but certainly I was beginning to understand the dimensions of the problem. Bone marrow transplants were being reserved for a small number of patients who would die without them. However, most of the transplant recipients were affected in other ways as well. They'd suffered serious organ damage and extensive burns. With this in mind, and having considered numerous scenarios over the previous few days, I suggested to Baranov the steps I would take to be of maximum service. He said I should present the proposal to Guskova, and we spoke with her together.

I began with personnel. Tissue typing was becoming very difficult because the patients' blood counts were falling. Moreover, large numbers of potential donors would have to be typed. One of the best people in the world in that area was Paul Terasaki. In some detail, I spelled out Paul's credentials for Baranov and Guskova. He was a professor of surgery, director of the UCLA tissue-typing laboratory, and president of the Transplantation Society, a much-respected worldwide organization with five thousand members. His greatest skill was in the area of automation. Instead of performing typing tasks one by one, Paul could get an assembly line rolling. He studied more patients in a day at UCLA than many of his colleagues study in a year, and he'd already prepared a mobile tissue-typing laboratory to bring to Moscow.

57

Dick Champlin was three years younger than I was, and my right-hand man at UCLA. Given the number of patients in Moscow, it was clear another clinical specialist was needed. Champlin would provide an additional pair of hands to perform bone marrow transplants and expertise in patient management.

The third area where I felt we needed assistance involved cases where perfectly matched donors were unavailable. In such instances, it would be necessary to remove the cells that caused graft-versus-host disease without destroying the cells necessary to regenerate new bone marrow. Here, the ideal person was Yair Reisner, who had developed a laboratory technique that made this possible. However, there was a problem. Reisner was Israeli, and diplomatic relations between the Soviet Union and Israel had been severed in 1967 after the Six Day War.

In asking for Reisner, I tried to make things more palatable by pointing out that he would be flying to the Soviet Union from New York. Still, I made it clear that he was an Israeli citizen who would be traveling on an Israeli passport. "All I ask," I told Guskova, "is that political differences be set aside to achieve the best possible care for your patients. We need someone with technical expertise who can perform the required cell separations. Yair is the ideal person for the job."

I also presented a list of supplies—various drugs and pieces of equipment—that I thought were necessary. Guskova said she would study the proposal and, with the afternoon at an end, Surik drove me back to the Sovietskaya.

After a short rest, I had dinner with Victor, Baranov, Fetisov, and Yevgeny Chazov—a Soviet doctor who had been Leonid Brezhnev's personal physician, was co-chairman of International Physicians for the Prevention of Nuclear War, and would eventually be named Minister of Health for the Soviet Union.

It was an elegant meal with lots of caviar, sturgeon, brandy, and vodka. We began by discussing friendship between our countries and the need to help each other, agreeing that certain things were beyond politics. Chazov spoke of the threat of nuclear war and, after several vodkas, I began to wax philosophic about how when I was a child I was amazed at how large the world was, but sitting now in Moscow, I realized it was very small.

Toward the end of dinner, Chazov mentioned that Russian Orthodox Easter services were being held that night and asked if I'd like to attend. I said yes. Several telephone calls were made and, despite the fact that reservations are required months in advance, our attendance was arranged.

Just before midnight, we arrived at a church in an old section of Moscow. It was an ornate building, brightly lit by candles, quite pretty and well-preserved despite its age. And it was packed, standing room only, with quite a few young people in attendance. Victor, in particular, was interested in the scene. He couldn't understand how fellow Soviets, particularly young men and women, could believe in religion.

We stayed for an hour, till one o'clock, then drove to a second church, where a large black limousine which belonged to the American Ambassador, Arthur Hartman, was parked outside. This second church was less imposing than the first, but just as full.

Back at the hotel, I began to unwind. It was two A.M., and I knew I'd be meeting Victor for breakfast at eight. Still, I wasn't ready to go to sleep. I read a little, listened to some Beethoven, played the events of the day just done over in my mind. And then, slowly, what was going on began to sink in.

I was the first Western physician invited to the Soviet Union to cope with a crisis since World War II. I had just seen the first victims of Chernobyl, and been witness to an atomic disaster. Something very important, and perhaps frightening, was happening.

CHAPTER

On Sunday, May 4th, I had breakfast with Victor. Then Surik came and drove us to Hospital Number 6. The previous day, we'd followed the flow of traffic, but that presented problems, since the streets were one-way and often quite long without a turnoff. This day, we discovered that by going the wrong way for a hundred yards down a set of trolley tracks, we could save ten minutes. It was the start of our efforts to beat the system, and led to a growing camaraderie between us. Thereafter, every morning, Surik would edge his way onto the tracks, look to see if a trolley was coming, and if not, speed ahead.

At the hospital, again we had a hard time getting past the babushkas. Each day, it would seem as if a different door was locked. One day, the front door would be sealed and we'd have to go around to the back. The next day, the back door was closed and we'd enter through the side. Getting to the sixth floor was also a challenge. I favored the stairs. Victor preferred taking an elevator but, except for the freight elevators which seldom ran, the elevators were tiny and held only two people each.

At nine A.M., the first of two daily patient-review meetings convened in the sixth-floor command headquarters. Baranov sat behind the desk, but Guskova, wearing a white surgical gown, was obviously the senior figure. Several other physicians were also in attendance, and as we discussed individual cases, it became clear that there were more patients than I'd thought. The previous day, sixteen names had been given to me. Baranov had mentioned "thirty patients" the night we met. Now it appeared there were twice that many. Young physicians came into the room periodically to report new data. At one point, someone brought us tea with chocolates. Then Baranov and I went to examine patients, the four I'd seen the previous day and several others. One was a middle-aged woman —a power plant security guard suffering from radiation burns. Another was a doctor named Orlov.

Of all the patients I saw in the Soviet Union, it was Dr. Orlov who affected me the most. A young man about my age, he'd gone to the reactor building after the explosion to help stabilize the condition of injured firefighters prior to their removal. As a doctor, he understood the nature of radiation. Yet he stayed in the reactor area for three hours, sacrificing himself to save others. Finally, feeling what he termed "a metal taste in the mouth and headache sickness," he retreated to the triage area. But even there, he continued his efforts.

When I first saw Orlov, he already bore signs of severe radiation sickness. Black herpes simplex blisters scarred his face, and his gums were raw with a white lacy look like Queen Anne's lace caused by candida infection. Then, over several days, the skin peeled away, and his gums turned fire-engine red like raw beef. Ulcers spread across his body. The membranes lining his intestines eroded, and he suffered bloody diarrhea. We administered morphine to ease the pain, but even when delirious, he remained in agony. The nature of radiation burns is that they get worse rather than better, because old cells die and young ones are unable to reproduce as a result of the damage. Toward the end, Orlov was barely recognizable, and his death several weeks after the disaster was merciful.

Late in the morning, Guskova told me that there was approval for Dick Champlin and Paul Terasaki to come to Moscow. I asked about Reisner, and she replied, "That must wait."

Knowing now that Dick and Paul would be coming, Baranov and I outlined what Terasaki would need in the way of laboratory space. Then we reviewed the catalogs of several Western medical-equipment and pharmaceutical companies, taking notes on what we'd need to augment our plans.

As on the previous day, it was decreed that I should go back to the hotel for lunch. That seemed like a fabulous waste of time and I told Victor that, first, I was happy not to eat lunch and, second, there must be someplace in the hospital where we could eat. After all, the patients ate lunch; the doctors and hospital staff couldn't possibly all go home to eat. But Victor said no, I'd been scheduled to go back to the hotel and that was inviolate.

It took us close to thirty minutes to get to the Sovietskaya. Despite the extensive menu in the hotel restaurant, the only entrée available was boiled beef, which was a shock to my system, since I'm used to frozen yogurt for lunch. Then, after we ate, Victor said it had been determined that I needed relaxation rather than more work, and I was taken on a tour of parks, churches, and other points of interest in Moscow.

Late in the afternoon, we returned to the hotel and I telephoned Los Angeles (where it was early morning) to launch Champlin and Terasaki. Then I called Hammer's office and spoke at length with Rick Jacobs regarding the equipment and drugs that were needed. Rick is exceptionally good at things like that. I simply said these are what we need, and we need them on Soviet electrical current. Rick had no idea what a Coulter Counter was—

or, to be more precise, a Coulter Counter S+4 Hystro-gram Differential With Reagents 500 SBC. Likewise, I doubt he knew whether a Fenwal leukopheresis machine was the size of a Sony Walkman or weighed a ton. But over the next few days, by telephoning medical-equipment manufacturers and pharmaceutical companies around the world, he got what we needed.

That night, I had dinner with Victor in the mezzanine restaurant. Downstairs, a loud band was playing as part of a protracted May Day celebration. Several of the songs were sung in English—"We Are the World" and "That's What Friends Are For." The menu listed dozens of items, virtually all of which were unavailable. We ordered, and were told our choices were out of stock. We ordered again with the same result. Finally, we asked for whatever was in the kitchen and they brought us dinner. It's no accident that Tolstoi and Dostoyevsky wrote thousand-page novels. They probably wrote them in Russian restaurants waiting to eat.

After the meal, Victor said we'd meet the following morning at eight for breakfast. Then I went upstairs to telephone Tamar and Hammer. A few reporters called my suite, and I told them each, "My obligation is to the patients; if I can help them, I will. That's what I'm here for, and speaking to the press could jeopardize everything. I hope you'll understand." For the most part, that was accepted by the media.

Monday, May 5th, began again with jogging, as did most of my days in Moscow. There were the usual negotiations, trying to get past the babushkas into the hospital, after which Victor and I went upstairs to the sixth-floor command headquarters.

Our morning meeting was chaired by Guskova. One by one we went over the patients, and as the review progressed, I found myself growing apprehensive as data were presented concerning patients I didn't know about. "Oh, he's new," someone would tell me.

From the beginning, I'd questioned whether the number of patients I'd been exposed to could account for all the soldiers and general activity at the hospital. Certainly, it had seemed likely that there were more patients, but I hadn't had a solid feeling for the data and wasn't sure Guskova and Baranov did either. Possibly, Baranov knew about the patients who had blood problems, but wasn't familiar with the ones who had burns without blood complications. Guskova might know who was in Hospital Number 6, but be unaware of new patients coming in from Kiev. In sum, I wasn't sure what they knew and how much of that they were telling me, and I was frustrated by not knowing the whole picture. Still, I trusted Baranov and Guskova and felt they were telling me what they could.

I did, however, make one suggestion. Each day, there was a constant flow of information, bits and pieces of data on different patients, and it was hard for me to keep everything straight. Partly, that was because the patients had long Russian names that were difficult for me to remember. But beyond that, the data were often presented in random fashion and not properly grouped. On the morning of May 5th, I suggested we make a large chart identifying each patient by number, and put all the information we had on each patient on the chart. That way, first the geneticist could come in and present data on each of his patients. Then someone else could tell us the results of tissue typing and the clinicians would give their report. Finally, when all the day's data were in, we could discuss the patients one at a time rather than jumping back and forth.

At the close of our meeting, Guskova approved the idea of a chart. Then Baranov told me that another bone marrow transplant was scheduled for that morning. This time, the recipient was a fireman named Varsinian.

Varsinian was a strong, young, very likeable fellow. Fighting the fire, he'd been exposed to a heavy dose of radiation and suffered severe hand burns from radioactive water. The Soviets have an antiseptic they put on burn areas which is ghoulish green. In the United States, it would never sell; it's the last color you'd choose. Whenever I saw Varsinian, he had green patches all over his body, and his hands were swathed in bandages.

Despite his wounds, Varsinian had a remarkably positive outlook and never complained. Given the nature of his injuries, it was difficult for him to demonstrate his burns, but he tried. And because he was in the sterile block, his cooperation was a real plus. Rather than have us enter the bubble, it reduced the chance of infection for us to stay outside and say, "Show us your hands; turn to the left." We'd ask Varsinian to lift his hospital gown and show us his leg, which was difficult for him to do because of the bandages, but he always managed. The only English he knew was "please" and "thank you." He spoke slowly but with a touch of playfulness, perhaps even with irony regarding his plight. A large tattoo covered his left shoulder, and I thought maybe he'd been a sailor before becoming a firefighter. I liked him enormously and thought he'd make it through the crisis.

As we had done two days earlier, Baranov and I drew bone marrow from Varsinian's donor together. Then we went back upstairs, where Guskova told me that Yair Reisner's coming to Moscow had been approved. I assume the decision was made at a fairly high level. Certainly, Guskova wouldn't have taken responsibility for it herself. But my sense is that if she'd argued against Reisner, he wouldn't have been invited.

That afternoon, Baranov and Guskova tended to patients who'd been in Hospital Number 6 before the Chernobyl accident. At their suggestion, I took several hours off, and someone from the Ministry of Health escorted me through the churches inside the Kremlin. Then I returned to the hotel and telephoned Reisner. He'd spent the previous few days gathering supplies at Sloan-Kettering and arranging to have things shipped to Moscow from his laboratory in Israel. Still, he was apprehensive about coming to the Soviet Union without a visa, and kept asking if I thought it was safe.

"Don't worry," I told him. "You'll be coming as a guest of the Soviet government."

That wasn't enough. Given the absence of diplomatic relations between Israel and the Soviet Union, Yair wasn't particularly keen on getting off a plane at Sheremetyevo Airport without knowing the language, not having a visa, and carrying only an Israeli passport. "I wouldn't go to Paris like that, let alone Moscow," he told me. Finally, I said that I personally would come to the airport to meet him. "All right," he responded. "But I absolutely refuse to get off the plane unless I see you."

Late that day, I went to the airport to greet Dick Champlin and Paul Terasaki. Paul had twenty-one luggage tags. Unfortunately, we only found twenty pieces of luggage. The twenty-first was a microscope, which was missing.

The Soviets had arranged for a truck to bring all the supplies to Hospital Number 6, and while we waited for it to be loaded, I gave Dick and Paul a brief rundown of what was happening in Moscow. They, in turn, filled me in on what they knew about Chernobyl from news reports in the United States. Ironically, although I was very much in the middle of things at Hospital Number 6, Chernobyl seemed quite distant. I had no knowledge about the accident, and couldn't see the forest for the trees. I knew even less about certain aspects of the disaster than the average Soviet citizen because I couldn't read *Pravda*. All I'd heard was that there had been an accident and that, in the Soviet phrase, the problem was being "liquidated."

We went first to the hospital, because Paul had brought a large number of antibodies and culture-typing trays that required refrigeration. Probably, we occupied every square inch of available freezer space. Then we spent several hours moving the rest of the equipment into place and, finally, went to the Sovietskaya for Dick and Paul to check in and eat dinner.

Obviously, the Soviets saw a hierarchy in things. Champlin and Terasaki had nice rooms, but not suites. Over dinner I realized how glad I was they'd come to Moscow. In addition to the medical skills they brought, it was nice to be with friends. Dick and I had known each other for eight years and co-authored numerous papers on our work at UCLA. Married with two daughters, he had a dry sense of humor and the ability to work near-endless hours.

Paul was a native of Los Angeles. Quiet and soft-spoken, with a voice that one strained to hear, he looked considerably younger than his fifty-seven years. His first experience with nuclear radiation had come in the 1950s, when he'd studied the effects of nuclear weapons tests, seeking out animals in the Nevada desert after blasts. During World War II, he and his family had been interned in Arizona for three years as a consequence of the Japanese "evacuation."

After dinner we retired to my suite and began what would become a nightly ritual of making plans for the day ahead. Usually, our sessions lasted two or three hours, and we worked on the assumption that whatever we said was being listened to by the Soviets via some sort of eavesdropping device. As a practical matter, we had nothing to hide. We weren't spies or planning to start World War III. And over time, we began to kid around that if we found any of the Soviets unpleasant to deal with, we could simply walk over to the wall in my suite and say, "You know, Ivan is really very friendly to us. I think he wants to become an American." And, of course, we speculated that if we did that, the next day Ivan would be thousands of miles away in Siberia.

At any rate, on our first night together I told Dick and Paul about my strategy for dealing with the media, that we could jeopardize our relationship with the Soviets and thus our ability to help patients if we spoke to reporters, and that I planned not to say anything for the time being. I suggested they follow this policy, and they agreed. Then

I told them a bit about Baranov and Guskova, how good they were, and emphasized that our role was to assist, and not displace, our Soviet counterparts.

Next, we began a more detailed discussion of individual patient cases. As the days passed, we wouldn't always be in agreement among ourselves regarding the best mode of treatment, but we did agree that it would be best to present the Soviets with one opinion. Thus, whatever our internal differences, we'd hammer them out each night in my suite and arrive at Hospital Number 6 in the morning with a single recommendation. Also, at our first night's session Paul made a very important suggestion. It was his view that once the crisis was over, rather than simply vanish, we should leave a legacy behind. In other words, we should teach our Soviet counterparts everything we knew, and give them the tools to improve transplant care in the Soviet Union after we'd gone. It was a good idea, and we tried to implement it.

The following morning after breakfast, Dick, Paul, and I went to the hospital with assorted drivers and translators. Terasaki met his Soviet counterpart, a woman named Ludmila Muzavieda, who understood tissue typing quite well but had been forced to labor with equipment that in some instances was forty years old.

Baranov was like a child on Christmas morning, opening boxes to see what Dick and Paul had brought. Many of the drugs, such as cyclosporine, were unavailable in the Soviet Union, and others were hard-to-get third-generation antibiotics. We spent several hours explaining dosage schedules, what the drugs were for, and how they should be monitored. Meanwhile, Victor was becoming more and more preoccupied as the airlift of medical equipment progressed.

Rick Jacobs had done his job well, gathering commitments for everything we needed. The Fenwal leukopheresis machines—there were three of them—were coming from France, Switzerland, and West Germany. Each one weighed a half-ton, and they were arriving on different flights, which necessitated quite a bit of trucking activity. The machines themselves were useless without plastic-ware, and that was coming from a warehouse in the eastern United States. Overall, twenty countries were involved in what would be a million-dollar airlift, and the movement of supplies around the world came to resemble a diagram from Oliver North's White House safe. Victor was responsible for seeing that everything got through customs and was delivered to the hospital without a hitch, and he performed that task with remarkable dedication.

Early in the afternoon, Dick, Paul, and I were driven back to the hotel for lunch. The previous day, I'd launched a successful mini-rebellion by packing lunch. Surveying the dining room at breakfast, I'd filled two bags with tomatoes, tongue, black bread, and fruit—one for me and one for Victor. Victor hadn't been pleased with the maneuver. He was into beefsteaks, not paper bags, for lunch. And also, possibly, he was concerned about violating the system. At any rate, on Dick and Paul's first day at the hospital, I was advised that new arrivals must be treated properly, and we ate at the Sovietskaya.

Yair Reisner arrived at six P.M. and, as promised, I was at the airport to meet him. Like Champlin and Terasaki, he'd brought dozens of boxes and, before doing anything else, we spent several hours at the hospital putting his things in order. Then a battle erupted.

Yair was born in Israel in 1948. I'd met him briefly in 1980, when he was studying at Stanford, but didn't get to know him well until 1982, when I spent a year at the Weizmann Institute. I was working in Israel on a project that entailed analyzing blood samples from patients with leukemia. Yair had a laboratory next to mine, and we frequently found ourselves working side by side late at night. That gave us ample opportunity to share ideas and get to know each other. What impressed me most was not Yair's technical brilliance, but his sense of responsibility. Whenever he does something, it's essential for him to do it right. He feels personally responsible for the outcome of every bone marrow transplant, even if factors beyond his control are at work. He keeps photographs of every patient whose bone marrow he's treated on the wall of his office.

Given this sense of responsibility, Yair was mortified by the centrifuge in his laboratory at Hospital Number 6. A centrifuge was necessary to separate blood cells that might cause graft-versus-host disease from those needed to restore normal bone marrow function. However, Yair's laboratory was on the fourth floor of Hospital Number 6, and the centrifuge there was badly antiquated. In response, the Soviets suggested he use a sixth-floor centrifuge adjacent to the sterile block, but Yair insisted that this was impossible. There was no way he could carry tubes between the fourth and sixth floors, because sterile conditions would be lost. He absolutely refused. The sixth-floor centrifuge had to be moved to his laboratory, period.

As the discussion progressed, I found myself dealing with a group of electricians and other hospital maintenance workers, all of whom were saying "This is impossible" and shaking their heads. They seemed to think that moving the centrifuge was a task akin to turning the Kremlin upside down. Meanwhile, Yair was growing increasingly excited, and Yair is someone whose emotions show on his face. Finally, I told the workers, "The sixth-floor centrifuge has to be moved."

"It can't be done. The door to the room is too small to take it out."

"If the door is too small, how did it get into the room in the first place?"

"Well, the machine was put in, and then the room was built around it."

Feeling very much between a rock and a hard place, I forged ahead. "Look, people's lives are at stake. Get the centrifuge out through the door or knock down the wall. The wall can always be rebuilt."

That didn't exactly make the maintenance workers

happy. There was much mumbling and grumbling, but the following morning when we got to the hospital, the centrifuge had been moved to the fourth floor where Yair wanted it.

Meanwhile, on the night of Yair's arrival, Dick, Paul, and I brought him into the fold at our nightly session in my suite. And by the following morning, Wednesday, May 7th, all of us were satisfied that things were moving on schedule. In particular, I was pleased with the evolution of the charts I'd suggested two days earlier. They were written in Russian but modeled on an American system, and the organization of data was now comprehensible. Pyatkin was coming in and giving us the results of tests for chromosome abnormalities. Paul and Ludmila were providing data on tissue typing. Slowly but surely, boxes were being filled in, updated, and revised where necessary.

Late in the morning, flush with victory from the previous night's centrifuge struggle, I decided to do battle with Lufthansa Airlines. As I mentioned earlier, when Dick and Paul arrived in Moscow, a microscope that was essential to Paul's tissue typing had been lost. For two days, the Russians had badgered the airline, but Lufthansa's people said they simply couldn't find it.

Unfortunately, no identical microscope was available in Moscow. This was an inverted phase microscope. Though not particularly valuable—it cost only about two thousand dollars—it wasn't standard, and we'd been forced to order a new one from the United States, which hadn't arrived yet. Meanwhile, the patients' blood counts were falling. Now, taking matters into my own hands, I telephoned a Lufthansa representative and said, "Look, we're in the middle of an international medical crisis. A million dollars' worth of equipment has been flown to Moscow. Four specialists have arrived from the United States and Israel. The eyes of the world are on us, and everything has come to a halt because Lufthansa lost a piece of equipment. Now I'm telling you, and I mean this seriously, in three hours I'm holding a press conference. Two thousand members of the world press corps will be there, and I'm going to tell them we can't do anything because of Lufthansa. Obviously, that will not be good for you, so I suggest you find the microscope."

Two hours later I got a call from the Lufthansa representative. He'd driven out to the airport himself and searched through piles of crates and boxes. He'd found the microscope and was on his way to the hospital with it. Please, would I not hold the press conference. That made an enormous impression on the Soviets. They were quite pleased with me and, more important, we were back in business.

My wife, Tamar, arrived in Moscow that night. The previous evening, with the Soviets' consent, I'd asked her to come. Dick, Paul, Yair, and I were away from the hotel for most of each day, and it was difficult for people to reach us. It occurred to me that if Tamar were at the Sovietskaya, she could process calls from the United States. Also, I missed her. Hammer had arranged for her airplane ticket, and Wednesday evening, for the third time in three days, I went out to the airport. Someone from the Ministry of Health had thoughtfully brought a small bouquet of flowers, which I presented with a flourish when Tamar stepped off the plane. Then we went back to the hotel, and that night had dinner as guests of Andrei Vorobiev and his wife.

Vorobiev was chief of the Department of Hematology at Moscow's Central Institute for Advanced Medical Studies and former director of the Department of Hematology at Hospital Number 6. A gracious man in his early sixties, he met daily with the Minister of Health and clearly was a senior player in the Chernobyl medical effort. During our meal, which was attended by several other Soviet officials and the usual entourage of translators, Vorobiev spoke at length on his view that men think they run the world but in reality it's women who make all the truly important decisions. At evening's end,

Tamar and I went back to the hotel, where, somewhat cantankerously, we spoke Hebrew in our suite, feeling it would make eavesdropping more difficult.

It was a time for reflection. Overall, things were going as well as could be expected. In particular, I was pleased by the mutual respect that was growing between Guskova, Baranov, and myself. We were dealing with Soviet patients in a Soviet hospital, and I was aware that they as Soviet doctors had final decision-making power. All I could do was make suggestions, but more and more they were soliciting my advice.

Baranov, in particular, impressed me with his knowledge. Even though he'd performed only a handful of bone marrow transplants in his career, he was well-informed and very competent. I'm familiar with virtually every article published in my field. Yet from time to time, he would refer to some obscure publication and I was surprised at how much he knew about transplant literature. More than once I thought it sad that Soviet physicians are isolated from the West. Baranov had an enormous contribution to make and, prior to Chernobyl, I'd never even heard of him.

I liked Baranov. He was warm, sincere, and extremely hard-working. Yair likened him to Israeli doctors of the 1950s—no big office, no sense of self-importance. If supplies had to be carried from one room to another, Baranov would pitch in and help the orderlies and nurses. In all our dealings, I don't think he ever misrepresented anything to me. I doubt he's capable of being dishonest. He was straightforward and, I think, both fulfilled and happy; a practical idealist with his feet on the ground and his head in the clouds. He was a good man, and that was important because difficult times lay ahead.

CHAPTER
7

By the end of my first week in Moscow, our daily routine was pretty much set. Each morning at nine, after I went jogging and ate breakfast, Dick, Paul, Yair, and I would meet with our Soviet counterparts at the hospital to discuss the patients. Then, through most of the day, we'd act on those discussions. As time went by, I met numerous relatives of the patients, many of whom had been flown to Moscow to be tissue-typed as possible bone marrow donors. Like most relatives in that sort of situation, they were hopeful, worried, and always seeking information. In addition to bone marrow transplants, a wide variety of treatments were utilized—platelet transfusions, chemotherapy, intravenous feeding, and the use of sophisticated antibiotics and anti-viral agents.

Generally, at five P.M., we'd reconvene on the sixth floor to review the day's developments. Baranov would record virtually everything that was said in a large black-and-white-speckled notebook. Guskova, although more removed from direct patient care, monitored the proceedings. More than Baranov, she seemed to have been shaped by hardship and, like him, she cared deeply for each patient. Whenever we went to examine someone together, no matter how serious the patient's condition, she'd always prompt me to say something encouraging. "We have a lot of women like her in Israel," Yair observed later. "Strong, tough, straightforward, and very decent."

Usually, we left the hospital around seven, had dinner at the hotel, and after that retired to my suite to discuss patients. When that meeting ended, I'd telephone Hammer and Rick Jacobs to give them an update and arrange for new medical-supply shipments. Generally, I didn't get to sleep until well after midnight. Throughout each day we were assisted by Victor and assorted other guides and translators. They were there to help us but, of course, their presence also meant that we were totally monitored.

Meals were our primary source of amusement. Each morning at breakfast, the waiters brought a large platter of sliced tongue, which we left uneaten. After a while the question arose: Was it different tongue each day or the same platter? Finally, Yair devised a method whereby we marked several slices with small incisions. The next day we checked to see if we got the same pieces. We didn't, although possibly someone else did.

Often, at dinner, we'd order beer when we sat down to eat. It wasn't served when they brought our borscht. It didn't come with the beefsteak either. Then, after we'd finished our ice cream, the waiter would appear with four bottles of beer, opened for each of us. Of course, by then we didn't want it. Beer is not a typical after-dinner drink. We'd take a few sips to be polite because beer is a precious commodity in the Soviet Union, but our hearts weren't in it. Then, finally, it dawned on us that the waiters were serving our beer after dessert precisely because we wouldn't drink it and, once the bottles were open, they could have it.

One of our most memorable meals involved Mikhail Bruk, a representative of Hammer's in Moscow. Bruk was Russian, but spoke near-flawless English. Depending on whom you talked with, he was variously identified as a KGB official, a high-ranking member of the Soviet Army, a higher-ranking member of the Communist Party, and just about anything else (good or bad) people could think of. Ask Bruk what his job was, and he'd say he was an editor for Novesti Press. What he edited, I don't know. But he was very well-connected.

Bruk was in his late forties and something of an aging macho yuppie. He had a house in Riga and an apartment in Moscow, drove a white BMW, and wore a diamond pinky ring and a shark's tooth on a gold chain around his neck. To say he stood out in Moscow would be something of an understatement. Victor met him once and was absolutely horrified.

Bruk and people like him exist because of a special tier in the Soviet economy. Most Americans have the image of an inefficient Soviet economic bureaucracy, with five-year plans, missed deadlines, and a bunch of people building tractors. It's known in Moscow that you don't buy a refrigerator or other major appliance made at the end of a month, because that's when factories are rushing to meet their quotas and the workmanship is incredibly shoddy.

However, there's another tier to the Soviet economy that's quite sophisticated and operates very efficiently. It's run by people who move around the world dealing in furs and negotiating for oil. They're not restricted by five-year plans or ideology, and they have considerable leeway in making decisions. Mikhail Bruk was such a person. The precise way he earned his salary, whether he was paid by Occidental Petroleum or the Soviet government, I don't know and doubt I could find out. Once Yair, Dick, Paul, Tamar, and I were settled in Moscow, he called and invited us out for dinner. We went, looking forward to the evening but aware that anything we said that night might be repeated and used to Mikhail's advantage.

Dinner was at the National Hotel restaurant, where the patrons were almost exclusively foreign businessmen and diplomats. Several Russian women were there, carrying Gucci bags and wearing up-to-date fashions from the West. The high point of the evening belonged to Yair. After several days in Moscow, he'd begun to feel comfortable and was no longer worried he might disappear. What concerned him most was Bruk's pretentiousness— the gold chains, the macho attitude. Bruk just wasn't Yair's kind of person. Midway through the meal, Bruk got up from the table and excused himself, saying he had to make a telephone call to Anatoly Dubrynin. As soon as he left, Yair went wild.

Yair is a brilliant mimic. He has a wonderful ear and sees things that make it easy to caricature another person. All of us notice certain traits in each other, but Yair re-creates and exaggerates just enough to make people come to life. And his sense of humor is spectacular. Ergo, Yair launched into a monologue, the premise of which was that Bruk had actually left to go to the bathroom but felt compelled to bolster his image with the mention of Dubrynin. Then, building on that, he had Bruk telephoning Mikhail Gorbachev and ordering Gorbachev to bring his BMW to the front of the restaurant to pick us up after dinner.

The day after our dinner with Mikhail Bruk, I went to visit the All Union Cardiological Institute. Yevgeny Chazov, whom I'd met my first full day in Moscow, was the institute's director and had extended an invitation for what was essentially a command appearance on my part. Dick Champlin and Tamar went with me. Chazov showed us around the facility at length and, as a gracious host, made the mistake of offering, "If there is anything I can possibly do to help the Chernobyl victims, tell me and I'll do it." I then amazed him by asking, "Would it be all right if we borrowed some equipment from your institute?"

In Moscow, medical institutes rarely lend equipment to one another. But there we were, in a very official setting

with translators around us, and having made the offer
Chazov couldn't very well turn around and answer no.
So he said what he had to say: "Why, certainly. Take
anything you want."

That began a fabulous search-and-seize mission. In every
laboratory we found something we wanted: centrifuge
tubes for Reisner, more tissue-typing trays for Terasaki.
By the end of the tour, our guides were laden down with
equipment and there were many unhappy Soviet doctors
with parts of their laboratories missing.

Meanwhile, the bone marrow transplants and other
forms of patient care continued. On May 8th, I was involved in my eighth and ninth transplants. The following
day, two more transplants were implemented. Then, on
May 10th, an odd incident occurred.

One of the Soviet patients, ironically also named Baranov, appeared not to have a matched donor. Accordingly,
once we ascertained his tissue type, I'd notified John
Goldman in London and John Hansen in Seattle and asked
them to search their computer banks in the United States
and Europe. It was a worldwide effort, and within forty-
eight hours Hansen reported a donor—a perfect match,
who was willing to fly to Moscow from New Orleans.
Then it turned out that the patient already had a donor
—his sister—and had asked her to leave the hospital,
partly because he didn't want her to be put under anesthesia and partly because he didn't want a transplant under any circumstances.

The risk in being a bone marrow donor is quite minor;
it's the risk of anesthesia. At most, one feels discomfort
from the punctures for about a week. I've given bone
marrow specimens for research purposes on more than a
hundred occasions. Still, I felt that we could not in good
conscience ask an unrelated person to donate bone marrow when the recipient had refused to let his own sister
take the risk. And given the fact that he didn't want the
operation anyway, we canceled the donor bank request.
In retrospect, the incident was enlightening in two re-

spects. First, it showed that the international donor-search network worked. And second, it was reassuring to know the Soviets weren't undertaking medical procedures without informed patient consent. Unfortunately, in this particular case, the patient was very wrong in his judgment. Lacking a transplant, he died several weeks after his exposure to radiation at Chernobyl.

On most days, Dick, Paul, and I left the hospital around seven P.M. However, Yair kept different hours. He couldn't begin to process bone marrow until it had been drawn from the donor, and often that didn't occur until late in the day, so his schedule was at odds with the rest of us. Just around the time we were going back to the hotel for dinner, Yair was starting work. Most nights, that kept him in the laboratory until after midnight, and his only companion was his translator, Vecheslov Stepanov.

If Yair and Mikhail Bruk were at opposite ends of a straight line, Stepanov could easily have been an equidistant point to form a triangle. Yair knows a lot about art and music, and is extremely well-read in European literature. Five-feet-seven and 150 pounds, he has a dark complexion and curly brown hair. Stepanov, by contrast, was tall, fat, and light-complexioned. Once a medical attaché at the Soviet Embassy in Washington, his hobby was collecting beer cans from around the world. And while superficially jovial, he wasn't particularly friendly. "Stepanov and I were alone in the laboratory for seven or eight hours at a time," Yair recalled later. "We had to talk. We were too bored not to." Still, according to Yair, Stepanov often turned the conversation toward Israel, asking very aggressively what it was that Israelis wanted.

In retrospect, one particularly moving incident involving Yair stands out in my mind. The Soviets had delayed several transplants waiting for his arrival in Moscow, and one of these was for a patient named Palamarchuk. Yair didn't finish processing the bone marrow from Palamarchuk's donor until well after midnight. Then, in the wee small hours of the morning, he and Baranov caucused. Baranov was wary of administering the bone marrow himself because he had never been involved with this type of laboratory procedure. Yair didn't want to do it because he was a biochemist, not a physician. So they did it together. Palamarchuk survived and, as of this writing, is healthy, the result of an Israeli scientist and a Soviet physician working together.

Still, despite our gains, we were starting to suffer losses. Rather than getting better, a number of patients were getting worse. They were growing disoriented, developing pulmonary problems, showing signs of major organ failure, and we didn't know why.

On Saturday, May 10th, three patients died—the first deaths since the day of the accident at Chernobyl. The deaths were expected. None of us thought we'd get by without victims. Still, it was sobering. We'd been operating in a very high-tech mode, wheeling supplies and medical specialists into Moscow, and now we had to face the reality that people were dying anyway.

I was with Baranov at a patient's bedside late in the day when the patient died. His name was Pravik. He was a fireman, married, with children. Pravik's wife was a nurse, and afterward, when Baranov sought to comfort her, she was crying softly. He patted her hand and spoke with her for several minutes. Then she left and Baranov, while professional and stoic, obviously felt the loss.

On May 11th, a fourth patient died, lowering morale a notch further. Meanwhile, the hospital staff was starting to fade. Many of the physicians had been on call twenty-four hours a day for fifteen days. They were physically exhausted and needed rest, but new patients kept pouring in. It now appeared that 299 plant workers and firefighters had been airlifted to Moscow. Two hundred of these had been exposed to substantial doses of radiation and were suffering from varying disorders which required intensive medical support.

By May 12th, the last of the bone marrow transplants

was complete. Some patients, who had been exposed to extraordinarily high amounts of radiation, were spared transplants because damage to other organs was such that they would probably die anyway. Other patients were treated satisfactorily with blood transfusions, antibiotics, and anti-viral agents.

On the afternoon of the 12th, the Soviets advised us that we were going on a forty-eight-hour "vacation" to Leningrad. With the initial stage of our work complete, they wanted to say thank you, and the plan was for us to leave Moscow by train at midnight. Paul Terasaki declined, choosing instead to leave for Paris, where he had a previously scheduled lecture to give. I refused, saying I didn't want to leave victims of the world's worst nuclear disaster to go sightseeing in Leningrad. However, Yair and Dick had finished their work, and there was no reason for them not to go. Tamar went with them, and once again I was alone in Moscow.

In their absence, Victor took me to the Tchaikovsky Theatre for a folk-dancing performance. Surik, our driver, had been working long hours, so we gave him the evening off and went by subway. It was the first time I'd traveled underground in Moscow, and the system was remarkable, particularly the subway stations, which were extraordinarily ornate. Tiled wall mosaics depicted revolutionary scenes from Soviet history. Marble pillars, vaulted ceilings, and chandeliers were much in view. In terms of glamour, the main stations resembled what New York's Grand Central Terminal must have looked like when it was first built. The fare was five kopeks, about seven cents. Everything was spotless.

After the performance, Victor and I walked back to the hotel. Then I went to sleep, generally satisfied with what we'd accomplished in the preceding two weeks, but knowing that formidable medical problems lay ahead. What I didn't foresee was how dramatically I was about to be thrust beyond medicine into an international political arena.

CHAPTER

All nuclear power plants have contingency plans for accidents. These include medical care, evacuation, fighting fires, and the like. In the United States, attention is also paid to dealing with the media. Most American electrical utilities have briefing rooms specifically reserved for use in the event of a nuclear accident. Even before a problem occurs, it's known who will be onstage, in what order they'll sit, and who will talk. All of this is designed to reassure the public, and while critical information is sometimes withheld, that withholding is vigorously challenged by the media.

There's no single correct way to handle disaster information. Reporters feel strongly that such information is in the public domain and that the process of gathering and disseminating it is as important as the end result. They want to be told what is certain and uncertain, and be able to describe events as they occur, not afterward. They want to be in someone's office when that person predicts that forty people will die, and then play back the tape a day later when the correct number turns out to be zero or eighty. I understand this process. An aggressive free press is essential to maintaining democracy. Yet at the other end of the spectrum, there are people who are interested in the end result, not the process. They feel that if you have two million citizens in Kiev and there's a nuclear accident fifty miles away, you should withhold information to avoid panic until confirmed data are available. Both sides want the public good, but they pursue it differently.

One of many dissimilarities between the American and Soviet systems is that if Chernobyl had happened in the United States, the hospital would have been swarming with reporters and there would have been press briefings every day. But as things were, I had isolated myself from the media in Moscow. I received periodic telephone calls from reporters, but didn't say much, and the Soviet government was largely silent. On April 30th, two days after announcing the accident, Radio Moscow broadcast a brief statement, saying that radiation levels in and around the power station had been reduced and work was continuing "to eliminate the consequences of the accident." Five days after that, a Tass dispatch updated the situation and reported the evacuation of Pripyat. But that, basically, was it. And as time passed, I began to feel that all of us involved in the medical rescue effort were losing credibility. I could justify saying nothing for a while, but eternal silence would lead inevitably to charges of a medical cover-up.

On Monday, May 12th, the day Tamar, Dick, and Yair left for Leningrad, I raised the subject with Victor. Essentially, I told him we had to say something to put an end to rumors and speculation, and that a press conference would be in the Soviet Union's best interests. I also pointed out that Dick, Paul, and Yair were readying to return home and, when they did, they'd be besieged by reporters. They couldn't be expected to remain silent forever. Nor could I.

Victor was not pleased to have this red-hot potato

dropped in his lap. But the following morning he reported back that a press conference had been approved. It was to be held two days hence, on May 15th, at the Ministry of Foreign Affairs, and I would conduct it in tandem with Andrei Vorobiev of the Central Institute for Advanced Medical Studies. Obviously, someone fairly high up in the Soviet hierarchy made the decision. And I'm sure they realized the ultimate issue was not whether or not a press conference would be held, but rather, when and in what country. Still, from the Soviets' point of view, agreeing to the press conference was something of a risk. They didn't know what I'd say, and I could only assume that, having dealt with me for two weeks, they trusted me.

Meanwhile, on the afternoon of May 13th, Hammer arrived in Moscow. His purpose in coming was to attend the opening of his art exhibit, "Five Hundred Years of Masterpieces," at the Moscow State Museum of Art. As always, he flew in on Occidental Petroleum's private jet, with an entourage that included his wife Frances; Rick Jacobs; Occidental's press officer, Frank Ashley; and Betty Arthur, a friend of Mrs. Hammer's. As soon as they arrived, Rick telephoned and said my presence was requested. Victor drove me to Hammer's apartment, and I went upstairs.

The apartment is a story in itself. For years, when Hammer and his wife traveled to Moscow, the Soviets put them up in the Lenin Suite at the National Hotel near Red Square. But Mrs. Hammer didn't like the accommodations. The plumbing was erratic, and too often the elevators didn't work. The rooms were stuffy because the windows were sealed. One night, when she and Hammer were readying for bed, she told him, "Armand, I've had enough of this place. I can't stand this hotel, and unless we can be more comfortable in Moscow, I'm not coming back."

The next night, Hammer attended a banquet at the Kremlin and was seated beside Andrei Alexandrov, a special assistant to Soviet Premier Leonid Brezhnev. Midway through the meal, Alexandrov turned to Hammer and said, "I understand Mrs. Hammer is not happy with your accommodations. Since you visit us so frequently, perhaps we could arrange for you to have an apartment in Moscow."

Several months later, when Dr. and Mrs. Hammer returned to the Soviet Union, they were given an apartment on Lavrushensky Street, across the Moscow River from the Kremlin. Their new accommodations had modern plumbing and electrical wiring. The furniture had been chosen by a West German decorating firm. Five families had lived in the space before the Hammers moved in.

Entering the apartment, I passed through a small foyer into the living room, which was large and comfortably furnished. A videocassette recorder and television were in full view. Clearly, though, the most impressive feature was the collection of elaborately framed paintings which hung on the walls. One in particular caught my eye. It was of two children, a boy and a girl, wearing peasant clothes. The girl was a bit older—maybe she was eight and the boy five—and in the painting they were exploring a wealthy merchant's mansion just after the revolution, staring in amazement at a piano. The point of it all was that in czarist times there had been such a dramatic separation between rich and poor that these children didn't know what they were looking at. They'd never seen a piano before.

The rest of the apartment consisted of a fully renovated kitchen, and a dining room, bedroom, and den. Most of the windows looked out on the Tretikov Gallery, one of the world's great art galleries.

Hammer greeted me warmly, and I brought him up to date on what was happening in Moscow—details of patient care; the planned press conference; Guskova, Baranov, and Victor, and what I understood their true roles and authority to be; how many transplants there had been; how many people were injured; how many had died. He wanted precise numbers and details on everything.

When I'd first arrived at the apartment, Hammer insisted that Victor stay downstairs. Finally, after about two hours, he called down and invited him up. Victor came upstairs, seated himself in the living room, and surveyed the scene—the television and VCR, the artwork, the size of the apartment. I'm not a mind reader, but Victor seemed to be thinking, "Well, with all its disadvantages, these are the benefits of capitalism."

The following morning, Wednesday, May 14th, I met with the Soviets to finalize plans for the press conference. This was serious business, and the atmosphere between us was quite formal. Victor, Fetisov, and Vorobiev were all there, as was Ivan Nikitin, chief of protocol for the Ministry of Health. The fact that the press conference was to be held at the Foreign Ministry suggested the Soviets viewed this as more than a matter of medical care.

The meeting began with my being told that the press conference would be jointly hosted by Vorobiev and myself. Each of us would make an opening statement and then, in rotation, answer written questions from reporters. I suggested that Dick Champlin and Yair Reisner be onstage with us, but that was vetoed on the grounds that then Baranov and Guskova would have to be onstage too and there'd be no end to it. Unspoken was the fact that the Soviets couldn't have been anxious to showcase an Israeli before the world press.

After our meeting, I went back to the hospital to visit patients. Then, that evening, in a twenty-five-minute nationally televised speech, Mikhail Gorbachev spoke publicly about Chernobyl for the first time. Describing the accident as very severe and of a magnitude never experienced in the history of nuclear power, he expressed sympathy for its victims and assured his nation that, "The worst is behind us." He further stated that details of the accident were released to the Soviet people and other nations as quickly as they were learned, and lambasted the Western media for "a mountain of lies" aimed at discrediting the Soviet government. All aspects of the problem, he concluded, were under study by a government commission and would be properly dealt with. Clearly, Chernobyl was more than a human tragedy. It had been an affront to Soviet honor, a national loss much like our own Challenger space shuttle disaster three months earlier.

Thursday, May 15th, dawned clear and bright. I'd wanted to prepare my opening remarks for the press conference the previous night, but had been too tired and gone to sleep instead. That necessitated getting up early Thursday and jotting down some thoughts on a lined yellow pad before I went jogging.

Tamar, Dick, and Yair returned to Moscow from Leningrad at six A.M., and we brought each other up to date on the preceding forty-eight hours. Then Victor met me at the Sovietskaya for breakfast. For the first time, as a consequence of Gorbachev's television appearance the previous night, the accident at Chernobyl was front-page news in Moscow. Victor translated the speech for me from *Pravda*, then asked for a copy of my opening remarks so they could be translated into Russian in advance of the press conference. The notes were handwritten and I gave him my only copy, assuming the handwriting was so bad that no one besides me could read it anyway.

At nine o'clock, the day's first crisis occurred. Hammer was planning to attend the press conference but wanted to visit Hospital Number 6 first. The night before, he'd requested permission, and the Soviets' answer was "nyet." Hammer was too old; he might get sick; it would be too much of a shock to his system. They had a hundred different reasons why they were very sorry but a trip to the hospital was impossible. Hammer's response was, "Hey, I'm a doctor. I've been working on this rescue mission for weeks. I want to see the patients."

At nine A.M., Rick Jacobs was dispatched to the hotel to enlist my support. I was scheduled to visit the hospital for a patient update before the press conference. Rick requested that, rather than go to the hospital, I go to Hammer's apartment, and then we could go to the hospital together. Hammer, Rick pointed out, had helped me get to Moscow. He'd worked like a dog and paid for the entire rescue mission. And Hammer wanted me at his apartment.

I knew if I left the Sovietskaya with Rick and went to Hammer's apartment, it would be the end of my relationship with the Soviets. I'd be siding against them, and they wouldn't like it. On the other hand, if I went directly to the hospital, Hammer would never get in and that would be the end of my relationship with Hammer. Clearly, I was between a rock and a hard place. Victor was saying, "Let's go to the hospital. This is what the Ministry wants, and the patients need you." Rick, meanwhile, was telephoning the Ministry of Health, Hammer was calling Anatoly Dubrynin, both sides were trying to cajole me into their respective limousines, Victor was re-checking with the Ministry, Rick was calling Hammer back to report on the battle, and Hammer was suggesting that someone explain the facts of life to me.

Finally, I said, "Okay, guys, I'll tell you what. I'm going into the hotel restaurant for a cup of coffee. You get this resolved and, when you do, come and get me."

My theory was that both sides knew there was a press conference and eventually they'd get me out of the restaurant one way or another. But as the morning passed, I found myself drinking quite a few cups of coffee with no sign of compromise. Finally, late in the morning, Hammer told the Soviets, "If you don't let me into Hospital Number Six after all I've done for you, I'll drive my car to the front of the hospital and stand outside until you let me in." And he would have. Hammer is a very stubborn man. And knowing that, the Soviets relented. Victor and Rick came into the restaurant where I was on what must have been my fifth cup of coffee and said they'd agreed I should go to the hospital with Victor. Rick would return to Hammer's apartment and drive him over to join us.

By the time I arrived at Hospital Number 6, Deputy Minister of Health Oleg Shepin was there with Guskova and the hospital director waiting for Hammer. They acted as though they'd been expecting him for months. Then Hammer arrived and we toured the facility for about thirty minutes, speaking with several patients and showing Hammer the areas where all the medical equipment we'd airlifted was in operation. Guskova gave him a small wooden bowl with salt in it, a traditional Russian gift. Then we got in our cars and drove to the Ministry of Foreign Affairs for the press conference.

Surik was driving; Victor and Fetisov were in the car with me. I'd decided I should go to the Ministry with the Soviets, since it was their press conference, and leave with Hammer. We arrived at the Ministry fifteen minutes late and were ushered into an anteroom adjacent to the main auditorium. Vorobiev was waiting with someone from the Foreign Ministry, who returned my opening-statement notes to me. Then I was told that there was a slight change. It was no longer a joint Soviet-American press conference. It was my press conference, and I'd invited Vorobiev to join me. That was fine, and we went over the ground rules again: an opening statement, now only by me; written questions for both of us—only written questions. At that point, for the first time, I said, "Only having written questions is unacceptable. I'll answer written questions, but the press is entitled to ask questions spontaneously too." That didn't exactly thrill the Soviets, but we compromised. The press conference would last two hours. Written questions would be answered first, and oral questions would follow if there was time remaining.

Then we walked out of the anteroom into the auditorium, and the crowd was extraordinary. I've faced large audiences before, but never like this. The entire world press corps seemed to be there. Camera crews, print media, radio commentators, photographers.

I sat at center stage, with Vorobiev to my right. At the last minute, the Soviets decided Hammer should join us, and a third chair was brought up and placed to Vorobiev's right. Then, more nervous than I thought I'd be, I began with a factual retelling of events—the number of people injured, how they'd been treated, and their prognosis for the future. I credited the Soviet physicians for being knowledgeable and involved. Prior to the press conference, I'd considered naming each of the companies and individuals from foreign countries who had supported the rescue mission, but decided instead to say simply that it had been an international effort. I did, however, single out Yair, Dick, and Paul. I was a little afraid that if I mentioned Israel by name, the Soviets wouldn't translate that portion of my remarks, so I identified Yair as being from the Weizmann Institute on the theory that many if not a majority of Russians would know the Weizmann Institute was in Israel. I tried to keep things as objective and straightforward as possible.

After my opening statement, I was given a pile of written questions and leafed through them. Basically, they fell into three categories. First, there were good questions. Some of these had been answered by my opening remarks, but the rest warranted attention, and Vorobiev and I responded to them. Next came trivial questions, such as how I liked the food in Moscow, and I ignored those on the theory that we had more important things to talk about. Last, there were polemical questions from both the East and West: "Don't you think this is a minor accident, since more people are killed every week by murderers in New York?" Or, "Isn't this accident worse than Hiroshima and Nagasaki, and a classic example of the Soviet Union's callous disregard for human life?" It was at this point that I discovered the usefulness of written questions. You don't have to answer them, and you don't have to give the questioners a forum for their diatribes.

After an hour, we opened the floor to oral questions—something that, to my knowledge, had never been done before in the Foreign Ministry. Hammer spoke briefly and drew applause with the announcement that all supplies and equipment airlifted to Moscow were his gift to the Soviets, free of charge. Vorobiev leaned toward dry statistical answers, except when describing the efforts of the firefighters, whom he justifiably regarded as heroes. At one point, there was some rumbling behind me, and I noticed Hammer getting up to leave. Then, a few minutes later, he returned, and Fetisov handed me a note saying, "You and Dr. Hammer will meet with Mikhail Gorbachev later today." Seconds later, speaking completely out of turn, an East German reporter grabbed the floor microphone and asked, "Have you any plans to meet with Mr. Gorbachev?" Obviously, it was a planted question, and I answered, "Yes, Dr. Hammer and I have been invited to meet with Mr. Gorbachev later today."

My feeling at the close of the press conference was that it had gone well. We'd given the media a chance to ask questions, defused some of the wilder rumors, and been fair and honest with everyone. If there was one moment that stood out in my mind, it came in response to a question regarding the lessons of Chernobyl. "Probably, I'll regret what I'm about to say," I answered. "But I think we have to view what's happened these past few weeks in a broader context. We've been dealing with a relatively small accident and, even with international cooperation, our ability to respond and care for the wounded has been limited. If we have a difficult time in helping three hundred victims, it's obvious that any response to the intentional use of nuclear weapons will be inadequate. People who believe meaningful medical assistance is possible for the victims of nuclear war are mistaken."

Following the press conference, Tamar and I went with Hammer to his apartment for lunch. Then, after he took a short nap, we drove to the Moscow State Museum for the opening of his art exhibit.

The gallery rooms provided by the Soviets were large and well-lit. The paintings themselves were well-positioned, and I know it made Hammer happy to see his private collection on display in Moscow.

Clearly, the top levels of Soviet society were in attendance. Everyone was well-dressed, and hand-held television cameras were much in evidence. Hammer spoke eloquently at the opening ceremony, sounding the theme that if the United States and Soviet Union could share their art, then surely we could develop a better understanding of each other's society.

Anatoly Dubrynin was at the opening, and I spent ten minutes chatting with him and Hammer and another half-hour with Tamar touring the exhibit. Then an aide to Dubrynin announced it was time to go to the Kremlin for our meeting with Gorbachev. We went downstairs, got into Dubrynin's limousine, and were off.

All the major streets in Moscow have a center lane reserved for government cars. Here, though, the entire road had been cleared, and we drove to the Kremlin without a single vehicle in front of us, passing block after block of cross streets with cars stopped and waiting. Hammer and Dubrynin were talking like old friends, which they are, and my mind wandered to an anecdote Victor had told me.

Both the Soviet Union and United States restrict the travel of each other's diplomats. For example, the Soviet Ambassador to the United States can't go to San Francisco. Thus, if he wants to travel to an area nearby, he has to fly somewhere else, get in a car, and drive. Similarly, the Soviets don't let our embassy staff travel to certain cities in the Soviet Union. These are the games that superpowers play.

Dubrynin was the Soviet Ambassador in Washington for years, and during that time he developed an intense liking for a particular pizzeria in Arlington, Virginia. However, in retaliation for the Soviets not giving our Ambassador a permit to travel someplace in the Ukraine that he wanted to visit, the State Department drew a gerrymandered map around Washington, something akin to a huge circle with a finger stuck in it. The purpose of this was to place the pizzeria off limits to Dubrynin.

Our drive to the Kremlin took about ten minutes. At the main entrance, we were met by two military attachés who led us to a small elevator, where a ranking Army officer greeted us. One of my peculiarities is that I wear wooden clogs instead of shoes whenever possible. I wear clogs in the operating room; I wear them at social functions. Whenever I have something on my feet, except for jogging shoes, it's wooden clogs. And I never wear socks with clogs. Now, I was readying to meet the leader of the Soviet Union, and the Army guide kept staring at my feet, looking exceedingly unhappy.

On the fourth floor, we got off the elevator and walked down a long corridor, where several dozen photographers were waiting. Gorbachev met us at the door to his office, shook hands with us, and ushered us inside. The photographers snapped pictures from an area just inside the door for several minutes and then were excused.

Obviously, there was a certain fascination with where I was—the Soviet equivalent to our Oval Office. The room itself was large but not grandiosely so, considering its occupant was perhaps the most powerful person in the world—more so than the President of the United States because there are fewer checks on his power. The carpet was green. The only painting I recall was a large portrait of Lenin behind the desk.

Gorbachev himself radiated power. It showed in his eyes. He's heavyset and of medium height, with a recognizable port-wine stain on his forehead. Port-wine stains are caused by dilated superficial blood vessels, which allow blood to be seen through the skin. They can be corrected surgically by removing the affected area, followed by a skin graft to cover the spot. In two or three weeks the area is completely healed, but obviously, cosmetic surgery didn't interest Gorbachev.

Gorbachev sat in a chair near the end of a long conference table facing the windows. Dubrynin was to his right, with Gorbachev's translator one seat further down. On our side of the table, Hammer sat opposite Gorbachev, while I was across from Dubrynin. Green felt mats, white lined pads, and round leather containers with pencils were on the table.

Gorbachev began by thanking Hammer and me for our help. He said he'd just come from a lengthy meeting of the Politburo, and although Chernobyl was a great tragedy, the Soviet people were very much behind him. This, of course, had to be put in context. At the time, Gorbachev was not considered particularly well-entrenched by the Western media, and political analysts were constantly asking, "Can he survive?" As if to answer the question, he reached for a soft leather folio and drew out several letters. One was from a factory worker in Odessa, whose fellow workers had voted to donate their next month's salary to victims of Chernobyl. Another was from a Soviet husband and wife, offering to shelter Chernobyl refugees in their home. The Soviet people, Gorbachev told us, had rallied to his side. Then he produced more correspondence, this time to show that support for the Soviet people in their time of crisis was worldwide. One letter was from a New York woman, who wrote that she wasn't as rich as Dr. Hammer but was enclosing five dollars to do her part. Another was from a woman, also living in the United States, who sent ten dollars, prompting Gorbachev to remark, "Apparently, she is wealthier than the first woman."

Then a look of anger crossed his eyes. "People all over the world are concerned with our situation and want to help. How is it that your government is acting so irresponsibly and takes pleasure in our tragedy?" By this, he was referring to American media reports. At one point,

United Press International had quoted sources as saying two thousand people were dead at Chernobyl. The *New York Post* had gone further, with a front-page headline that shrieked, "Mass Grave: 15,000 Reported Buried in Nuke Disposal Site."

"Dr. Gale," Gorbachev said, "you know the extent of the injuries. How can your government act like this?"

In the United States, of course, the government doesn't control the press. Gorbachev is an intelligent man and should understand that. However, there's a large gap between intellectual understanding and the actual appreciation of something when you've grown up with the exact opposite. If you're fifty-plus years old and, like Gorbachev, all you've seen your entire life is a government-run press, the converse is probably hard to accept.

Hammer responded, saying, "It's unfortunate that these articles appeared, but in our country the press is free to print what it chooses. For better or worse, the government does not control it."

That did nothing to placate Gorbachev. Secretary of State George Shultz, he pointed out, had charged the Kremlin with deliberately understating casualties "by a good measure." President Reagan had engaged in similar anti-Soviet propaganda. "This mountain of lies," Gorbachev continued, "is designed to poison the hearts of the world against the Soviet people, and divert attention from warlike American actions in the Persian Gulf and Gulf of Sidra."

Gorbachev was angry. It wasn't an act. His face was red and, as he spoke, he repeatedly jabbed the point of a pencil into a felt mat in front of him. When my turn came, I said simply that no American took pleasure in the misfortune of innocent people; that in the absence of hard data, the media had misreported some facts, but certainly my presence in Moscow was an indication that Americans cared and were willing to set aside political differences to help the Soviets. That didn't satisfy Gorbachev, but it was enough to move him to the next subject.

"You know, your President is pursuing a course of action that will put the world further at risk. What would happen if Chernobyl had been in space? What if nuclear weapons exploded in space by mistake?" This was an emotional argument rather than a scientific one. At present, no one is suggesting putting nuclear weapons in satellites. However, it served as a bridge to Gorbachev's next theme: "If you put your Strategic Defense Initiative system into space, we will do likewise. We don't want to, but we'll have no choice."

Looking at the history of superpower negotiations—not just between the United States and Soviet Union, but between all countries—communications often flow through alternative channels. Someone makes a speech at the United Nations; the Hungarian or Swiss Ambassador is called in and given a message; a businessman is told something and reports back to officials. These alternatives are important, particularly when diplomatic relations have been broken or aren't going well between two countries. Sitting with Hammer, Dubrynin, and Gorbachev, I realized that this wasn't simply a congratulatory meeting, where someone gives you a tie clip and an autograph. This was a diplomatic exchange at a fairly high level, a world I wasn't used to traveling through.

Hammer responded that he agreed with Gorbachev that both sides should do everything possible to prevent the accidental detonation of nuclear weapons. However, he didn't think it was his role to judge the propriety of particular weapons systems. By now, Gorbachev's anger had abated slightly, and Hammer took it upon himself to move the discussion in a different direction.

"Mr. Secretary General, the only way I know to reduce the threat of nuclear weapons is to have a dialogue between our countries. Secretary of State Shultz and Foreign Minister Shevardnadze were supposed to meet several months ago, and that meeting was canceled by the Soviet Union after the American bombing of Libya. Mr. Shevardnadze is going to London in the near future. Perhaps

Mr. Shultz could meet him there and lay the groundwork for a future summit. You met President Reagan at Geneva, and you liked each other. And, of course, President Reagan is very popular with the American people. It's not easy for a president to negotiate arms reductions, but Mr. Reagan's base of support is sufficiently strong that he can make concessions another president might not be able to make and still win treaty ratification in the United States Senate."

It was a bold offering, but Gorbachev was unconvinced. "Yes, I met Mr. Reagan in Geneva," he answered. "That meeting served the purpose of our becoming acquainted, but I'm not interested in meeting again just to say hello. Something productive must be accomplished."

Undaunted, Hammer continued to press. "Mr. Secretary General, you're the most capable, realistic Soviet leader since Lenin, and the one most likely to achieve peace. The greatest thing you could do is come to Washington and meet with Mr. Reagan."

Still, Gorbachev was unmoved. "Your administration acts as though I must go to Washington for a summit, but that isn't the case. I refuse to meet with Mr. Reagan on his terms just to meet."

The conversation continued in that vein until, finally, the hour allotted for our meeting had elapsed. On signal, we rose to leave and then, just before we reached the door, Hammer did something that surprised me. "Mr. Secretary General," he said, stopping and turning back, "you have a lot of Jewish people in the Soviet Union, and many of them want to leave. If someday you reach an agreement with President Reagan, the United States Senate would be much more likely to ratify it if the plight of the Jews were to change. These people don't want to be in your country. They're giving you a hard time and a lot of bad press. Why don't you let the Jewish people go?"

Gorbachev seemed taken aback by the question, but recovered quickly. "Dr. Hammer, Jewish citizens like it in the Soviet Union. In many countries, they are discrim-

inated against, but here there is no anti-Semitism. Our
Jewish people are happy. Whatever else you hear is Western propaganda."

Then we shook hands, and Hammer offered one final
thought: "Premier Gorbachev, there have been two great
desires in my life. One is to find a cure for cancer. The
other is to see world peace. I'm eighty-eight years old and
don't have much time left. I hope that you and Mr. Reagan
will come to an agreement."

Gorbachev smiled. "I'm a young man," he answered.
"I can wait longer for peace."

After the meeting, Dubrynin's chauffeur drove us back
to Hammer's apartment. Hammer was exhausted and went
to sleep. Meanwhile, I went with Rick Jacobs to Occidental Petroleum's Moscow office, where he summoned
a secretary and I did my best to reconstruct our session
with Gorbachev.

Occidental's offices were fairly modern, consisting of
a large waiting area, several conference rooms, and Hammer's personal office, which had a picture window over-
looking the Moscow River. By the time I finished the
reconstruction, it was early evening and another secretary
came into the conference room to tell us that
"Vremya"—Moscow's equivalent to our network news
—was about to start. That morning, my name had been
mentioned in *Pravda* as a consequence of Gorbachev's
speech the previous night, so the Soviet people knew
American doctors were aiding the relief effort. Overall,
though, I'd been relatively anonymous.

Now, suddenly, everything changed. Chernobyl dom-
inated the entire news broadcast, starting with a report
on the press conference, then an interlude for Hammer's
art exhibition, and finally our meeting with Gorbachev.
I sat in a conference room watching the telecast, and
when "Vremya" ended, the entire press conference was
broadcast.

Television is a potent medium, and I question whether

any private American citizen ever received as much TV exposure in the Soviet Union as I did that night. When the broadcast was over, I was driven in an Occidental limousine back to the hotel. Then not fully realizing the implications of what had happened, I went to sleep.

The following morning, Friday, May 16th, I readied to leave Moscow. Three days earlier, I'd decided I should to go back to Los Angeles. Our immediate work at Hospital Number 6 was coming to a close. And while extensive medical care for the transplant patients would be required for several months, we were in a period where most of the patients were progressing nicely. Meanwhile, there were a number of things I had to do at UCLA.

On the day Hammer arrived in Moscow, he and I had agreed I'd go back to Los Angeles for ten days and then return to the Soviet Union. Dick and Yair would leave with me, but their return was unnecessary. Instead, Dick would cover for me at UCLA for much of the next year, and Yair would concentrate on his work at the Weizmann Institute.

On the morning of May 16th, I went to Hospital Number 6 for a last visit with Guskova and Baranov. Both of them were wearing the white coats I was used to seeing them in, and I told them what a wonderful experience it had been to work with them. Guskova responded, "We are a team, and I hope the relationship continues. You and your colleagues have done everything possible to help our patients. We deeply appreciate it." We shook hands, and Baranov walked me downstairs to the car outside the hospital. "Thanks a lot," he said.

From there I went to Hammer's apartment and, after that, with Hammer to the Ministry of Health. At the Ministry, Oleg Shepin presented me with a nineteenth-century Russian painting of an old wooden church in a Siberian snowstorm as a gift from the Soviet people. Then we drove to the airport, where Hammer's plane was waiting.

Oxy One is a white Boeing 727, fitted with extra fuel

tanks to give it a range of 4,600 miles. The crew consists of two pilots, an engineer, and two stewards. Pursuant to Soviet air regulations, when Hammer flew to Moscow, the Soviets sent a third pilot and navigator to meet him at Heathrow Airport in London. The Soviet airmen didn't fly the plane, but were required to be in the cockpit.

We boarded from the rear. On the right was the first of three bathrooms, wood-paneled and carpeted, with a large mirror and marble-topped sink with gold fixtures. Then came the galley, with a microwave oven, coffee-maker, and refrigerator stocked with fresh fruit, cheeses, caviar, and other delicacies. After that came a lounge area with sofas and chairs convertible for sleeping, and, further down on the left, Hammer's private suite with an office, bedroom, and bath. The front half of the plane had a second lounge with two more sofas and a table that could be raised and opened into a dining setting for eight. Then a small room with a desk, copy machine, and large world map, followed by the crew's quarters and cockpit. Hammer likes movies, so there were a large videotape collection and two VCRs onboard; also a compact-disc player and numerous discs. A sophisticated communications system allowed passengers to make telephone calls from the plane to any place on Earth.

Hammer, his traveling party, Yair, Tamar, and I were the only passengers. Dick Champlin would take a regularly scheduled flight to Los Angeles a day later. The doors closed, the plane taxied down the runway, and I have to say, taking off was a fabulous feeling. The Soviets couldn't have been nicer; they hadn't stopped me from doing anything I wanted to do. But being on Oxy One meant I was going home.

The flight to London lasted three hours. At Heathrow Airport, we dropped off the Soviet navigator and pilot, and said good-bye to Yair, who flew from there on an El Al flight to Israel. Then the plane took off again, the crew converted the sofas into beds, and I slept until we reached Los Angeles.

There was a short press conference at the Los Angeles airport, after which Tamar and I went home to see the kids. Overall, I was satisfied with the journey, but I knew there'd be little time to rest. After an initial honeymoon period, the weeks following a bone marrow transplant are filled with life-threatening dilemmas. Most deaths occur not immediately, but thirty to ninety days after transplant.

Serious problems, and considerable international political intrigue, lay ahead.

PART
THREE

The Journey
Continues

CHAPTER

Since the spring of 1986, I've been to the Soviet Union six times. Each trip has taught me something about the world we live in, and each trip has led me to further survey my life to discover who I am.

On my father's side, I can trace my roots to my great-grandfather, who brought his wife and children to the United States from Byelorussia. One of his children was my grandfather, who worked most of his life as a tailor for a garment manufacturer in New York. My father was born in 1908 and grew up in Brooklyn. His name was Harvey Galinsky, but while employed by the Metropolitan Life Insurance Company, he was advised by a supervisor that a "simpler" name might facilitate getting ahead in the organization. Thus, he became Harvey Gale, before leaving Metropolitan to take a backroom job with a Wall Street brokerage firm.

My mother was several years younger than my father and came from a comparable background. However, her brothers and sister became financially quite successful. One brother, Ted, went to Hollywood in the 1930s, appeared in a number of movies, and counted Mae West and Cary Grant among his friends. Adolph, my mother's other brother, made a fortune in women's high fashion as an entrepreneur. Min, my mother's sister, married a prominent lawyer who died young, and her second husband was an even wealthier attorney. So each of my mother's siblings was well-off; two of them were multimillionaires. And at times I sensed that left my mother feeling somewhat out of sorts.

I was born in New York on October 11, 1945, and grew up in Flatbush, a predominantly Jewish, middle-class section of Brooklyn. We lived in an apartment at 75 East 21st Street. My younger brother, Steven, and I shared one bedroom; my parents slept in the other. I went to Elementary School 249 and Junior High 246, then high school at Erasmus Hall.

Erasmus is the second-oldest high school in the United States. Only Boston Latin in Massachusetts is older. During the Revolutionary War, the Battle of Brooklyn was fought on what became the school's grounds, and Alexander Hamilton and John Jay once sat on its board of trustees. In 1895, the academy was given to Brooklyn for use as a free public school, and since then several hundred thousand students have passed through its doors.

I was a successful student in high school, channeled into programs for the gifted, and my grades were good. Athletically, I was too small for football, and played soccer instead. On weekends I went to Pratt Institute for classes in industrial art. Occasionally, I frequented Greenwich Village coffee houses, listening to poets like Allen Ginsberg as my one counterculture indulgence. I also developed a strong affinity for opera.

Both of my parents were devoted to the cultural aspects of New York. My father told me that the first time he went to an opera it was because my mother dragged him to *La Bohème*, and he was absolutely captivated; he couldn't get over it. Before long, he knew everything the Metropolitan Opera performed by heart. Unlike my mother, who went both for the performance and social sideshow, my father was interested in the performance alone. Whether he was in standing room or the dress circle made little difference. All he cared about was that it be the best performance possible.

Some of my earliest memories are of my parents studying the newspaper to see what was happening at Carnegie Hall. They went into Manhattan two or three times a week, going to free events as well as expensive ones. Often, they brought me with them, and I remember taking the subway home late at night after evenings I'm not sure they could really afford. Still, their enthusiasm took hold, and to this day I appreciate good music and art.

During my senior year of high school, I worked for the *New York Times,* which was in competition with the *Herald Tribune* to have its papers distributed in social studies classes. The average paperboy delivers fifty papers to fifty homes daily. At Erasmus, there were eight thousand students and whoever had the *Times* job delivered two thousand papers to a handful of classrooms. I shared the contract with another student. Each morning, we got to school between six-thirty and seven. For ten hours of work per week, I made eight or nine dollars an hour, most of which was saved for college. By the end of each day, I was pretty tired. I'd gotten to school to schlep papers while most of my classmates were still in bed, and four afternoons a week I stayed late to play soccer. Still, it was worth the effort.

I dated very little; I guess that was shyness on my part. Each year, there was a crisis as New Year's Eve approached, but my parents were close friends with a couple who had a very nice daughter. Usually, around three o'clock on the afternoon of December 31st, I'd ask my mother to call this girl's mother to find out if she had a date. Fortunately, she'd be free, and I'd invite her out.

Overall, I was self-governed. My parents had uncompromising standards when it came to schoolwork, but I never felt as though they were pressuring me. I knew they were there if I needed help and that I had their support. Both of them were dedicated to giving their children what they thought was important.

I graduated from high school in 1962 and, at age sixteen, went off to college. That much seemed preordained. Growing up, it had been clear that I was on a professional track. Both of my parents had started college but never finished—although ultimately my mother returned to get her degree when she was fifty years old. In my case, the question wasn't whether but where. My mother wanted me to stay near home, and argued that the New York City university system was superior to most private colleges. But my father was absolutely determined that I get out of Brooklyn. His view was that the exposure I'd had through high school was limited to one city, and that life in New York wasn't typical of life in America. I wasn't particularly well-oriented toward colleges. I didn't know which were "good" and which were "out," which were prestigious, and where it was easy or hard to get accepted. Initially, I leaned toward Lehigh because of an interest in engineering, but fate intervened in a rather bizarre manner.

At the time, there was a very popular quiz show on television called "General Electric College Bowl." The format matched students from different schools against each other, and during my senior year of high school, Hobart College won several contests. One of the members of Hobart's team lived in our apartment building—at least his parents did—and this made a huge impression on my father. Also, Hobart was out of the city, in upstate New York. I applied for admission, the admissions office seemed happy to have me, and that was that.

So I enrolled at Hobart, an Episcopal liberal arts college in Geneva, New York. The entire student body numbered 1,200 students, two-thirds the size of my graduating high-school class. And it was a good experience for me. I'd come from a very homogeneous group of people, and at Hobart maybe 10 percent of the student body was from New York City. I joined a fraternity, Sigma Chi, which was exclusively Christian on campus until I pledged. Again, I tried out for the football team, and again I didn't make it, so I played lacrosse instead. I did well academically and broadened my intellectual interests, moving beyond science to study philosophy, literature, and other aspects of Western civilization. I grew a beard, let my hair grow long, lived for a while in the fraternity house, and learned a lot about people who I hadn't known existed when I was growing up in Brooklyn.

The decision to go to medical school evolved during my junior year of college. In high school, I'd wanted to be a nuclear physicist. A cousin, ten years older than I and very bright, was a nuclear physicist and used to tell me wonderful stories about his work. It seemed that this was where the action was, that nuclear physics would have an enormous impact on global issues, philosophy, even mathematics. Then, in college, I turned toward the biological sciences. Physics, and particularly theoretical physics, seemed too remote. I could see the tangible impact of what I was doing in biology. I don't recall giving it a lot of thought, but medical school became the logical direction for me.

Neither of my parents was particularly keen on medicine. In fact, one recollection I have is of my mother saying under no circumstances should I become a dentist because dentists have to put their hands in other people's mouths. That seemed repulsive to her, but one night, years later, I was in an emergency room as an intern treating a gunshot victim and found myself drenched in blood and excreta. At that moment, I remembered my mother's warning: "Don't be a dentist; you'll have to put your hands in other people's mouths."

As with college, choosing a medical school was fairly simple. The State University of New York at Buffalo had a policy of admitting students as it interviewed them rather than making them wait. I was interviewed early in my senior year. They offered me a spot; all of my peers were living in fear of not being accepted by a medical school, so I accepted. At the time, I had no reservations about not holding out for a more prestigious place like Harvard or Columbia. My knowledge of the players was very naive; I didn't know what the competition was. And my basic philosophy was and remains that these things are relatively unimportant. I think I've done as well coming from Hobart and Buffalo as I would have coming from Columbia and Harvard.

Medical school to me was about academic discipline, learning to think and solve problems through careful deductive reasoning. The first two years consisted of classes. The next two were clinical; we began working with patients in the wards, and this was where I excelled. Even today, my major strength is one of synthesis. I can take divergent data in complex medical situations and make sense of them. By the end of my third year of medical school, I was functioning as a de facto intern and doing well.

During the summer of 1968, after my second year of medical school, I took my first trip out of the United States—to Gondar in northern Ethiopia. As a second-year student, I'd had a wonderful idea. Ethiopians have a very low incidence of heart disease, which struck me as interesting. I wondered why, and managed to convince the United Health Fund to give me a fellowship to study abroad. Needless to say, I felt quite important. I set off from the United States with my portable EKG machine, arrived in Gondar, and discovered they didn't have electricity, which rendered both the machine and my entire project inoperable.

At that point, I abandoned my research and spent three months traversing the countryside with local health-care officials. Gondar was nowhere, and starting from there we'd travel for days by donkey to villages where people had never seen a white man before. They didn't know they were in Ethiopia. They didn't know Ethiopia was in Africa. They knew their land to the horizon and nothing more. There were a lot of lepers, smallpox victims, and other desperately ill people. Many times, especially at night, I wished I were home. But one of the wonderful things about being a doctor is you get to help people. In one village, the chief's daughter, a young girl, was ill with pneumonia. We gave her penicillin, she recovered, and the chief was so happy he slaughtered a cow for a feast in our honor. Throughout the countryside, we dispensed "miracle" drugs. To cure someone, to take children who are dying and make them well—there's nothing more gratifying a person can do.

Ethiopia was the start of my medical wanderings. The following summer, I spent three months in Thailand, where a very different health-care system prevailed. In Ethiopia, there had been virtually no doctors, and people with a high school education were trained to care for patients and administer drugs. By contrast, Thailand's medical system was a product of the Ford Foundation. Promising students were given fellowships to study in the United States and then returned to their homeland as doctors. Unfortunately, though, most refused to go into the countryside. They insisted on staying in Bangkok, and medical services beyond the capital were primitive and limited.

In 1970, after four years of medical school, I began an internship. I wanted to go into academic medicine; that is, serve on a university faculty and conduct research in addition to patient care. However, selecting a hospital for my internship required considerable deliberation. Some hospitals have relatively few patients with extensive supervision, and doctors are expected to know everything there is to know about each case, no matter how complex. At the other extreme, there are hospitals where interns learn by doing, supervision is relatively light, and you can't possibly know everything about every patient and disease because you're trying frantically just to keep people alive. I preferred the former environment, feeling I could learn enough from two or three gunshot wounds. I didn't need to see a hundred of them and, with that in mind, ultimately I pursued my internship at UCLA.

The UCLA Medical Center is among the foremost medical research institutes in the world. Its primary objective is medical education, although patient care is excellent and it's one of eighteen comprehensive cancer treatment centers in the United States. The typical course a young doctor follows is to be an intern for a year and a hospital resident for two. Then some doctors go into practice on their own, but most select subspecialty training and stay on for another year.

In my case, the process went astray. One of the foremost immunologists in the country—John Fahey—had come to UCLA in 1970. Fahey was an M.D., but at UCLA served as head of the Department of Microbiology and Immunology, which was not a clinical department. He needed a physician to work with; I was looking for something interesting to do, and we decided that after a year of residency, I'd stop my formal medical training to enroll as a Ph.D. candidate in microbiology and immunology. Thereafter, I spent several years doing basic work with mice, hoping it would lead to treatment of human cancers by manipulating the immune system. Then, quite by chance, I fell into bone marrow transplantation.

The National Institute of Health had given UCLA a huge machine designed to separate blood into its various components. The way it worked was, you hooked a person up to the machine; blood was removed from a vein in one arm, processed, and returned to a vein in the other arm. Today, it's a common process, but our machine was an early model, and we had no idea how to use it. For about six months, it sat in a corridor outside our office staring at us and vice versa. Finally, a rather eminent film producer came to us with a medical problem. We thought we could cure him immunologically by extracting blood cells, treating them, and returning the blood to his body, but that required getting this machine working. A manufacturer's representative visited the hospital and showed us how to use it. Then I went to the blood bank, got several bags of expired blood, and constructed a pseudo-human to put blood through the machine to see if I could work it. Everything was in order, but bags of week-old blood are different from a live patient. Common sense dictated that I do my first work on a real person at the side of an experienced teacher. Meanwhile, two similar machines were in operation—one in Texas and one in Seattle. Since Seattle was closer, I called the hospital there, asked for help, and they said, "Come on up. We use it every day." So with Fahey's permission, I went to Seattle, where for no apparent reason they decided not to use the machine for a week. That left me with a lot of free time, and at one point I asked, "What do you do with this thing, anyway?"

"Bone marrow transplants," they answered.

At the time, I knew nothing about bone marrow transplantation. The first successful transplant hadn't even been performed until 1969, and there had been over two hundred failures before that one success. But with nothing else to do for a week, I figured I might as well learn about the process. And very quickly, by observing what was going on in Seattle, I realized that all the resources necessary for bone marrow transplantation were present at UCLA.

Again by chance, during this period a hematologist named Martin Cline came to the UCLA Medical Center. He was interested in starting a transplant program, and soon I found myself spending half my time as a Ph.D. student with John Fahey and the other half with Cline developing a bone marrow transplant program. I never completed my residency, but finished my post-doctoral medical training by applying laboratory work toward the required experience. At one point, after joining the medical faculty, I had to resign in order to receive my Ph.D. degree. Otherwise, I would have been awarding myself a doctorate. But in the end, everything fell into place. Today, I'm an Associate Professor of Medicine. Prior to Chernobyl, I spent roughly 40 percent of my time in teaching and patient care, 40 percent in the laboratory, and 20 percent doing administrative work. Most of the latter involves managing the program's million-dollar annual budget. Money comes largely from university funds, government contracts, and grants. I never trained to be a corporate executive, but the job requires it.

That, in a nutshell, is my "curriculum vitae." I've been married twice—once in the early 1970s to a psychiatric social worker who's now an attorney in Los Angeles, and later to Tamar, whom I met traveling in Israel. We were married in 1976, and in more ways than I can express, she's my Rock of Gibraltar.

I take pride in what I do professionally, and only once that I'm aware of have my actions been called into question. That was in 1978, when a series of bone marrow transplants were performed at UCLA. Each of the physicians involved decided a transplant was the preferred treatment for his or her patient, obtained the patient's permission, and proceeded. Then the National Institute of Health launched an investigation.

Within the past decade, most hospitals have established federally mandated institutional review boards to determine which experimental treatments are appropriate for use on consenting patients. Physicians practicing outside of hospitals don't need committee approval. They can recommend any accepted treatment provided it's for the patient's benefit and with patient consent. However, in a hospital setting, there's a fine line between what's "experimental" and what isn't. If a doctor says, "These ten patients need bone marrow transplants, it's the only way they'll live," all he needs is patient consent. However, if the same doctor says, "I wonder what the results will be if we perform bone marrow transplants on the next ten patients," that's experimental. The key is motive —whether the treatment is chosen specifically and completely for the patient's condition or in part for research purposes. In the latter instance, even with patient consent, the physician must submit his study to the review board for approval.

After the 1978 transplants, the National Institute of

Health concluded that they had been "experimental" and that, because the physicians who administered them reported to me, I was responsible. In its finding, the NIH acknowledged no harm had been done. It emphasized that it was reviewing only the experimental nature of the treatment, not its efficacy or the skill of the physicians involved. It continued to retain me as a consultant and to award me NIH grants. Nonetheless, I was formally admonished.

Clearly, someone performing a medical experiment should get approval from an authority more knowledgeable than the patient. However, to this day, I don't think the 1978 transplants were experimental. Rather, the issue was one of performing intensive medicine to save lives without being uncaring. Certainly, transplantation techniques have advanced since then. If the same bone marrow transplants were performed today—and two thousand of them are annually—they wouldn't be considered experimental. Yet, by the same token, in many ways, transplantation is still in its technical infancy. Years from now, what we're doing in 1988 will seem as crude an "experiment" as putting leeches on people to cure them by bleeding. Also, I have to say, if a patient were to die because I hadn't done everything in my power to save him, the pain would be worse than a formal admonition ever could be.

Back in the United States after our return from Moscow, I had a week to get my life in order before going back to the Soviet Union. On Saturday, May 17th, the day after we arrived home, I went to UCLA to review the condition of my patients. Dr. Winston Ho had covered for Dick Champlin and myself during our absence, and everything seemed to be in order. There was a short news conference at the medical center to discuss our trip, and afterward an office party where we demolished a one-kilo tin of caviar I'd brought back from the Soviet Union.

On Sunday, May 18th, Tamar and I attended Hammer's

eighty-eighth birthday celebration in the Grand Ballroom of the Beverly Wilshire Hotel. It was quite a production. The black-tie dinner began at seven P.M. and was accompanied by a series of performances. Mstislav Rostropovich played the cello, accompanied by Jean Barr on piano. Ludmila Lopukhova and Simon Dow performed the "Pas de Deux" from *Swan Lake*, Act II. Merv Griffin served as master of ceremonies, and there were toasts by Gregory Peck, Los Angeles Mayor Tom Bradley, Abigail Van Buren, and the Reverend Robert Schuller. Miniature grand pianos served as rotating centerpieces on each table. Among the guests were Charlton Heston, Alan Cranston, Betsy Bloomingdale, Edmund Brown, and James Roosevelt.

Three days later, I was in Hammer's company again; this time, on a trip to Washington, D.C., to meet with Secretary of State George Shultz. We left Los Angeles around noon on May 21st, flying onboard Oxy One. Dick Champlin, Paul Terasaki, and several Occidental personnel flew with us. At Dulles International Airport, we were met by company limousines and driven to the Madison Hotel, where I checked in under an assumed name. That was Hammer's doing.

The next morning at nine, Dick, Paul, and I went to a private showing of the Soviet's Impressionist exhibit at the National Gallery. Hammer, you'll recall, had been in Washington for the opening the day I learned about Chernobyl, and he arranged for an assistant curator to guide us through the gallery that morning before it opened to the public. One painting, in particular, caught my eye— a Van Gogh oil with prisoners walking in a circle in a prison yard. It captured the American stereotype of a Soviet gulag completely. We also visited a National Gallery collection of war photographs dating from Crimea, which was the first war after the invention of photography. Two pictures were especially moving to me: the first, a twisted watchtower at Hiroshima after the blast; the second, a photograph of American military personnel sitting with their wives on the deck of an aircraft carrier watching an atomic test in the South Pacific.

At one P.M., Hammer and I went to the State Department to meet with George Shultz. Dick, Paul, and Rick Jacobs came with us. Someone from Shultz's staff met us in the lobby and brought us upstairs to a waiting room where Ambassador Hartman was sitting. Hartman and I chatted briefly. I don't think he'd been recalled to Washington for this meeting, but rather had been there on other business and was asked to sit in.

Then Shultz appeared in the doorway and ushered us inside. He's a large man, powerfully built, at least six feet tall. His office was the same size as Gorbachev's, but much less formal. One area had a fireplace, stuffed armchairs, and colonial American antiques. It looked more like a living room than an office, and that's where we sat.

Shultz was friendly but quite serious. He began by saying he'd spoken with President Reagan, and wanted to convey the President's appreciation for everything we'd done to help the Soviet citizens injured at Chernobyl; that our actions reflected well on the American people. Hammer responded that we'd been happy to help, and suggested I provide a more detailed account of what occurred than had been available through the media. I spent about twenty minutes recounting my experiences, and Shultz posed a number of questions. Had I been to Chernobyl? What was the state of Soviet medical technology? At one point, he digressed to say the United States had made it clear from the start that we wanted to help the Soviets, and that there had been no "gloating" over their plight. Any exaggerated casualty reports appearing in the media were the result of the Soviets' withholding information and the fact that, in the United States, the media are free to say what they want. This led to a discussion of our meeting with Gorbachev, and here Hammer took the lead, suggesting there must be a way for both sides to capitalize on Chernobyl as a road to nuclear arms reduction and a more stable peace.

Toward the end of the meeting, Hammer showed Secretary Shultz several photographs I'd taken of patients in Hospital Number 6. Shultz then thanked us and asked if I'd seen the official State Department rooms one floor above. I hadn't, and he instructed an aide to take Dick, Paul, and me to view various banquet rooms and historical desks, while he spent another half-hour alone with Hammer. After the meeting, we were hungry, and I asked the limousine driver to stop at a frozen yogurt stand for a modified "power lunch." Then, late in the day, we flew back to Los Angeles.

Overall, I was impressed by Shultz. He's one of the few people in the forefront of Reagan administration foreign policy whom I regard as a top-notch professional. Our meeting had been arranged in part to relay a report on our exchange with Gorbachev, but also because Hammer wanted to make it clear that we weren't trying to circumvent the United States government in any way. Obviously, the State Department would have preferred to be more in control of the Chernobyl relief effort, and just as obviously, it doesn't like private citizens carrying out personal dealings with the Soviet government. But I think we impressed Shultz as responsible individuals rather than dangerous instrumentalities who were likely to go out of control.

Still, one incident seemed odd to me. Normally, Hammer would have had an Occidental photographer at our meeting with Shultz. That's the way he operates; a company photographer records everything. However, the State Department made it clear it wanted its own photographer, and that Occidental's photographer wasn't welcome. That's understandable. But during the meeting, the official photographer didn't seem to be paying attention. He wasn't focusing or aiming his camera. At best, he was just shooting. Later, the State Department told us its pictures hadn't come out. There were no photos at all of Shultz with myself and Hammer.

I don't know if that was intentional or not. One assumes official State Department photographers know how to use a camera. But it's possible Shultz didn't fully trust us; that, despite being cordial, he wasn't comfortable with who I was or what we were doing in the Soviet Union. And if that's the case, he might not have wanted photographs. If photos had been released, they would have been on page one of a lot of newspapers. Robert Gale and Armand Hammer, just back from the Soviet Union, meeting with Secretary of State Shultz. What would that have meant? Would it have been interpreted as a government endorsement of our meeting with Gorbachev? Something more? I didn't know what to make of it.

Back in Los Angeles, I had two days to prepare for my return to Moscow. Dick Champlin and Winston Ho agreed to cover for me at UCLA while I was gone. Several dozen of my daughters' classmates at Roscomare Road School gave me letters for delivery to Gorbachev. "I hope you are having a nice time and doing things right," wrote one student. "I want our countries to have peace and love." "I hope no more people die," wrote another. Many of the letters were adorned with crayon drawings of hearts and flowers. One nine-year-old, who appeared more advanced than the others, wrote, "Hello, Mr. Gorbachev, how are you? I think you might put a containment structure around the reactor so an accident won't happen again. You also might want to install safety rods and an emergency cooling system."

The final twenty-four hours before departing again for Moscow were spent gathering items I'd been asked to bring back to the Soviet Union. Victor needed a battery charger. His daughter wanted frozen pizza. Baranov requested an infinite quantity of Polaroid film to photograph patients. If an American goes overseas to Europe or Asia, no one says, "Bring me back a lightbulb or tube of toothpaste." But every Soviet citizen has a shopping list of items from America. I gathered them all as best I could, although the frozen pizza was deferred to a later trip.

Then, on the afternoon of Saturday, May 24th, I returned to Los Angeles International Airport for my next journey to Moscow.

CHAPTER

10

I arrived in Moscow for my second Soviet trip on the same Lufthansa flight I'd taken twenty-three days earlier. Victor and Baranov met me at the airport, and we drove to the Sovietskaya. This time, though, there was a different feeling. I was coming back to people I trusted and places I knew. As before, I offered to go to the hospital immediately, but Baranov said, "No, you're tired. Rest and we'll go tomorrow." In point of fact, I wasn't tired at all. I'd slept for nine hours on the plane, but timing and formality are very important in dealing with the Soviets.

On Monday morning, May 26th, I went to the hospital. During my absence, there had been several deaths, and Guskova asked me to evaluate the sickest patients first. Many of them were approaching a critical period which would determine whether they lived or died. A few were patients I hadn't seen before—people previously thought to be suffering only from thermal burns, who had been in a different ward. Some patients we had anticipated would do well weren't. Fortunately, the converse also was true.

One thing I noticed was that the patients were beginning to regard me differently. Partly, that was because they were used to dealing with me. But beyond that, their relatives had reported seeing me on television and reading about me in the newspapers; Mikhail Gorbachev had sung my praises, so I had a new stature in their eyes. One of the patients, a fireman named Tarmosian, who had received a bone marrow transplant during my first trip, actually complained bitterly that I'd deserted him by going back to Los Angeles.

At the end of day one, I distributed the booty I'd brought from California—Polaroid film and a hematology textbook for Baranov, silver earrings for Guskova, a battery charger for Victor. I'd also brought something for the patients—copies of the catalog from the National Gallery's Soviet art exhibit. It was an attractive book with color reproductions and proclamations from both Reagan and Gorbachev.

Meanwhile, Baranov and I were getting to know each other more intimately. During my first trip, his wife had been in the hospital with a slipped disc, and he'd taken time every day to visit her. Now she was better, and I felt his relief. We began to talk about our families—he had two grown sons—and his English seemed to be improving daily, although at most meetings Victor still acted as translator.

Clearly, Baranov was determined to do everything possible to help the Chernobyl victims. He was also a very complex man with a desire to succeed. More than once, he was late for dinner or left a hospital session early to attend a Communist Party meeting. In the United States, medicine is largely divorced from politics. There are instances, notably the present AIDS and abortion controversies, where regrettably the two coincide. But for the most part, aside from hospital or university "politics," the two spheres are separate. Intelligence, skill, and ambition are far more important to success in our medical profession than whether a person is a Democrat or Republican.

By contrast, in the Soviet Union, someone who wants to get ahead goes to Communist Party meetings. Waiters do it; truck drivers do it. It's power, life, ideology, and religion. I asked Baranov once who went to his meetings, and he told me it was physicians who worked in the hospital. They discussed their work in an ideological context. I guess that raises the question, why doesn't everyone in the Soviet Union join the party? After all, there are only about eight million members, despite the fact that membership is considered essential to success. The answer, I suppose, is it takes time to go to meetings, they're probably not very interesting, and some people are simply ideologically opposed.

During the last week of May, three more patients died, including the woman plant guard who had received a fetal liver transplant during my first trip to Moscow. Each death was hard on the surviving patients. They were from the same plant crews, the same firefighting units, and suffering from the same illnesses. More and more time was spent going to autopsies and reviewing slides from tissue specimens, trying to determine precise causes of death. Still, for every loss, there were signs of hope, with three firemen holding special promise. Varsinian, Tarmosian, and Palamarchuk had all received bone marrow transplants during my first trip. Each was responding well to treatment and appeared to be on the road to recovery.

On Sunday, June 1st, after a week in Moscow, I flew to Kiev. For some time, I'd been troubled by the feeling that I was in an information vacuum. Back in the United States after my first trip to the Soviet Union, I'd learned as much as I could about Chernobyl. But each new piece of information raised as many questions as it answered, and people kept asking, "What's happening in Kiev? Is there food to eat? Are the people drinking radioactive water?" I also had a more personal reason for going. For over a month, I'd been living, breathing, and sleeping Chernobyl. I wanted to know what it was all about. I didn't understand the relationship of the power plant to the town of Pripyat, which areas had been evacuated, how many people had been treated medically, and so on. I wanted to see what conditions were like in Kiev, fifty miles from the accident.

Finally, I told Victor it was important for me to go to Kiev and, if possible, Chernobyl. Every step with the Soviets is a negotiation. Getting to Kiev required convincing them that my going would help them, and my argument was, "Look, I'm being asked questions. I want to speak intelligently about your handling of the situation, but I can't support or defend you without firm data. By now, you people should trust me. So if your government has been telling the truth, it's in your best interests for me to visit the affected areas."

On Friday, May 30th, Victor reported back that the appropriate authorities had approved my going. Two days later, he, Nikolai Fetisov, and I took a commercial flight to Kiev. The plane was jammed, indicating that things couldn't be too dangerous at our destination. Dr. Mikhail Shandala led a delegation that met us at the airport, and by late afternoon I was motoring through the Ukrainian countryside on a modern expressway. The drive took about thirty minutes. First, we passed through suburbs, then across a bridge over the Dnieper River into Kiev, the third largest city in the Soviet Union, with two and a half million residents. The entrance to Kiev is singularly dramatic. Just past the river there's a huge statue, larger than our own Statue of Liberty—"Mother Russia." Painted silver, it's of a woman holding a shield in one hand and a sword in the other, a monument both to victory and to the twenty million Soviets who perished in World War II.

The streets of Kiev were tree-lined and hilly, reminding me of San Francisco. People were more colorfully dressed than those in Moscow, and I was fascinated by a novel traffic-control device that was much in evidence. Most lights were attached to a digital timer that counted down from the moment the light turned red or green so drivers would know how much time they had before the color would change again. Everything on the streets seemed normal except for one thing—there were no children.

At the time of the Chernobyl explosion, the wind had been blowing northeast toward Scandinavia rather than south toward Kiev, and the force of the explosion resulted in radioactive material being concentrated at an altitude of five thousand feet. Thus, even though Kiev was relatively near the disaster site, it was spared many early consequences of the accident. But on May 5th, nine days after the explosion, the wind changed and radiation levels in Kiev rose to eighty times normal. Residents were advised to keep all windows closed, shake the dust off their clothes after being outdoors, shower daily, and only eat food from stores where radiation levels were inspected regularly. Then, in mid-May, the children of Kiev had been evacuated to summer camps, where they would remain until August as a precautionary measure. Their absence was a telling reminder of the dangers of Chernobyl, and if another were necessary, it soon became clear that Kiev must have been the cleanest city in the world because its streets and buildings were being washed daily. Wherever I looked, people were hosing down everything around them. Obviously, the radioactive particles being washed away would flow into the sewage system and, ultimately, the Dnieper River. But the first priority of the Soviet government was to halt the concentration of radioactive material on the streets and buildings by diluting it in water. It was a matter of degree rather than absolutes.

Victor, Fetisov, and I checked into the Kiev Hotel, where we had dinner with Anatoly Romanyenko, Minister of Health for the Ukraine. Then we were escorted to the Ukrainian Medical Museum, which was impressively automated with television sets, dioramas, moving walls, and other electronic devices. Basically, the point the Soviets were making with this visit was that their society was responsible for innumerable contributions to medicine. For example, I was asked if I knew who the first person was to describe the symptoms of a heart attack in a living person. I answered "James Herrick," which is what most American physicians would say, since that's what we've been taught in medical school. But I was told no, it was a Soviet doctor, ten years before Herrick. Whether that's true or not, I don't know. The museum also devoted considerable space to Soviet physicians who cared for the wounded in World War II, and Shandala made a point of showing me a photograph of Armand Hammer that was part of an exhibit depicting the great famine and typhus epidemic of the early 1920s.

The following morning, Monday, June 2nd, began with breakfast at the hotel. Then I was taken to Hospital Number 14, the major municipal hospital in Kiev, and spent several hours being introduced to patients. All totaled, several hundred people had been screened there in the aftermath of Chernobyl, but almost immediately the most seriously injured of these had been airlifted to Moscow and the rest had long since been released.

Next on our tour was the Ukrainian Radiological and Cancer Institute. Here, there had been fewer patients than were once in Hospital Number 14, but their injuries were more serious, and several had died before the airlift to Moscow was put into effect. Two of the patients, still undergoing treatment, were the first off-site victims of Chernobyl I'd encountered. One was a woman employed as a cook at the power station. She'd woken up on the morning of the disaster and gone to work without knowing of the explosion. How that could happen, I don't know, but in the process, she'd walked across a field of tall grass coated with radioactive particles and suffered serious leg burns. The second patient was a man who'd heard about the accident and rode his bicycle to the power station to see what was happening. As he rode, the bike kicked up radioactive dust and, wearing sandals, he'd received burns on his lower legs and feet.

The last hospital we visited was the Kiev Institute of Pediatrics, Obstetrics and Gynecology, where I spoke at length with Yelena Lukiyanova, the institute's director, about pregnant women who'd been exposed to radiation. A fetus is most vulnerable to mental retardation as a consequence of radiation during the eighth through fifteenth weeks of gestation. That's when cells that have previously clustered together migrate out to form the adult brain. Worst-case scenarios indicated an increase of up to 50 percent in the birth of retarded children in the danger zone around Chernobyl.

Lukiyanova told me that all pregnancies in Kiev were being carefully monitored. There had been a meeting of the Ukrainian Academy of Science to set up guidelines for abortions in the aftermath of the accident. However, Academy members couldn't reach an agreement, other than to acknowledge that different women had different interests in bearing children. For example, a woman with six children who conceived quite easily and had been exposed to a large dose of radiation might elect to have an abortion, whereas a childless woman who had great difficulty becoming pregnant might want to carry her child to term. The final decision reached by the Academy was that all obstetricians should be educated regarding the dangers of radiation in order to explain the situation to their patients, but that there would be no uniform policy or guidelines to govern abortion decisions.

After the third hospital visit, we went back to the hotel for lunch, and by this point a fairly large entourage had joined us. As a consequence of media exposure, a lot of Soviets knew my name and face, and people in Kiev saw me as something of a hero because I'd willingly joined them on the perimeter of the danger zone. That made for a rather hectic lunch, but by then I had a greater concern. I was scheduled to take a helicopter to Chernobyl that afternoon, but suddenly people from the Ministry of Health were appearing, vanishing, and appearing again. Then Shandala told me the weather was quite bad around Chernobyl. Rainstorms were forecast. It was too dangerous to go—because of the weather. That might have been true. But it was also possible the Soviets just didn't want me at Chernobyl. In other words, bring me to Kiev, show me what they wanted me to see, and send me back to Moscow.

When Victor (who had vanished during the proceedings) reappeared, I told him, "You know, Victor, if I go back to Moscow without seeing Chernobyl, people will think something horrendous is happening, that the situation still isn't under control. You wouldn't want that, would you?"

More telephone calls were made. Possibly, Victor himself didn't know the true status of things. But in any event, finally, he came back and reported, "Not today, maybe tomorrow." That left the rest of the afternoon free, and I told him I wanted to go to Babi Yar.

Babi Yar is a valley that no longer exists, the site of the worst single massacre of World War II. Literally translated, it means "Grandmother's Forest." I won't go to Germany without trying to visit Dachau, and there was no way I could be in Kiev without paying tribute to the victims of Babi Yar. Outside of the concentration camps, it's the greatest horror story of the war. Over one hundred thousand Soviets, primarily Jews who lived in Kiev, were told by occupying German troops that they were being evacuated and should bring all their belongings to Babi Yar. There was a ravine with corrals and tunnels leading to the precipice. Machine-gunners were waiting in place on the opposite side, and as the victims approached, they were gunned down, falling forward into the ravine. One hundred thousand men, women, and children. They must have heard the fire from the guns, but were trapped and had no place to go. "Most of the dead were poor and illiterate," D. M. Thomas later wrote. "But every single one of them had dreamed dreams, seen visions, and had amazing experiences; even the babes in arms; perhaps especially the babes in arms."

When the Germans realized they were going to lose the war, they went back to Babi Yar and tried to burn the bodies, but there were too many of them. Later, the Soviets filled in the ravine and built a monument on a grassy knoll. It depicts steps leading to a precipice, with sculpted people climbing upward, a few at the peak, and still more past the pinnacle falling forward. To the Soviets, Babi Yar wasn't a massacre of Jews. It was an act of genocide against the Soviet people on Soviet soil.

That night, Victor and Fetisov took me to a concert. Cultural events are well-publicized in the Soviet Union and, whenever possible, I went to see them rather than simply going back to my hotel room. As a rule, each morning Victor would ask what I was interested in seeing and produce tickets later in the day. The Bolshoi Ballet, recitals at Tchaikovsky Hall, opera at the Large Theatre within the Kremlin walls—I saw them all. Invariably, the audience was well-educated about music, but what impressed me most was a ritual I saw. At the end of each piece, people would walk up to the stage and present flowers to the performers. These weren't roses thrown from the orchestra pit or store-bought bouquets an usher would bring onstage. Rather, they were single flowers from home gardens, personal thanks from the audience to performers who had given them pleasure.

After the concert, Fetisov turned in for the night, while Victor and I stayed up for several vodkas. By this time, he and I were becoming more open with each other. He even admitted missing things about the United States, although overall professing happiness at being back in the Soviet Union. One of Victor's few regrets about his stay in Washington was that his daughter hadn't attended an American school. She was thirteen years old at the time, and he'd felt she could benefit from American schooling, but he and his wife were afraid it would be dangerous. There was "too much anti-Soviet feeling in Washington, too many strange groups running around," so they sent her to the Soviet compound school instead.

One of the things I'd learned that afternoon was that milk from dairy farms around Kiev was showing excess levels of radioactive iodine as a result of Chernobyl. However, the half-life of iodine is only eight days, and in several months the milk would be "harmless." Thus, the Soviets were turning it into butter and cheese. That, of course, left open the question of who would eat previously radioactive butter. But the vodka was good, and after a while Victor and I decided that his comrades could learn from American capitalism. "Make it a collector's item," I told him. "Market the butter in small lead tubs as butter that glows in the dark or self-melting butter. Or better yet, tell people all they have to do is spread some of the butter on frozen bread and it will turn to toast."

In truth, Victor understood the advantages of capitalism. He was loyal to his country, but he'd seen America and wanted the Soviet Union to evolve economically. Economic inefficiency, government bureaucracy, a babushka who wouldn't let him through a hospital door—those things drove him crazy. He was well aware that the Soviet economy isn't motivated by personal incentive to the extent that it should be.

Victor himself worked very hard. In a society where physicians and truck drivers make approximately the same salary, where there's little correlation between productivity and rewards, he drove himself mercilessly. I guess he exemplifies that part of human nature independent of economic systems; someone with a work ethic who derives satisfaction from meeting challenges and doing his job well.

The following morning, Tuesday, June 3rd, Victor told me at breakfast that we were going to Chernobyl. Shortly after eight, we got in a limousine—Fetisov and Shandala were in another—and drove to an airport on the outskirts of Kiev. No one at the airport seemed to know what was happening. There were telephone calls, a fifteen-minute wait, and again I was unsure if the trip was on or off. Finally, we were ushered past the gates to a helicopter with room for one passenger in front and three more in back. Victor, Fetisov, and Shandala got in back, and I was given the seat beside the pilot. We put on helmets with a fairly complex walkie-talkie system enabling us to talk over the roar of the copter but not easily. To speak, you pushed one button; to listen, another. It was impossible to do both simultaneously. Several bottles of mineral water and cups were onboard in case any of us got thirsty. We strapped on our seat belts, and the copter took off.

It was a misty, foggy morning, and each step of the way I found it hard to believe I was actually going to Chernobyl. The pilot had been there many times before as a volunteer after the explosion. Because of the weather, it was difficult to see the ground, and I imagined someone announcing at any minute that we couldn't go on. I wasn't sure whether Victor, Fetisov, and Shandala were coming because they wanted to or out of a sense of obligation.

Slowly, the fog lifted. We flew for an hour over marshlands with beautiful rivers and streams. I was taking mental notes, memorizing as much as I could, but not writing anything down. I'd brought a camera, but wasn't going to take photos until we reached the power station. A flock of geese flew beneath us. Then we entered the evacuation zone.

In point of fact, the power station was ten miles from the town of Chernobyl. The city of Pripyat was much closer. Pripyat was a company town, built virtually overnight in the early 1970s to house engineers, carpenters, plumbers, and other construction personnel at the station site. As construction was completed, the first set of workers moved out and plant operating personnel moved in. By April 1986, Pripyat had 49,000 residents, four schools, ten restaurants, an arts center, sports complex, movie theater, dance hall, and twenty-three apartment buildings. Then Chernobyl Unit Number 4 exploded.

Within thirty minutes of the accident, the plant medical station had been alerted, and a team of local physicians began administering on-site treatment. By six A.M., less than five hours later, 108 firemen and plant personnel had been admitted to regional hospitals for decontamination, emergency trauma treatment, and evaluation of radiation exposure. Dr. Guskova had arrived at noon, as dictated by a plan developed in the late 1960s to deal with nuclear accidents. Pursuant to this program, the most seriously injured were sent to Hospital Number 6 in Moscow, while others were treated in Kiev and various regional hospitals. All totaled, approximately 100,000 people were examined in the days immediately following the accident. It was an extraordinary effort requiring the services of twelve hundred doctors, nine hundred nurses, three thousand physician's assistants, and seven hundred medical students working in shifts to provide twenty-four-hour-a-day coverage. Eighteen thousand individuals were hospitalized, and it's a tribute to Soviet planning that only thirty-one people died in the aftermath of the explosion.

Meanwhile, ninety minutes after the accident, Major-General Gennadi Berdov arrived in Pripyat. Summoning police and militia members from nearby towns, he sealed off the reactor complex and erected roadblocks throughout the area. Residents of Pripyat were instructed to stay indoors and keep their windows tightly shut. All schools and commercial establishments were closed. Later in the day, a task force under the command of Boris Shcherbina, Deputy Chairman of the Soviet Council of Ministers, arrived and took control. Checkpoints were established along the sixty-seven-mile perimeter of a "forbidden zone," and plans to evacuate everyone within six miles of the power station were formulated.

That night, eleven hundred buses were driven to Pripyat from Kiev. At 1:50 P.M. the following afternoon, thirty-six hours after the accident, residents were summoned to building stairwells and told they were being evacuated. Ten minutes later, the exodus began. By 4:20 P.M., Pripyat was empty, its 49,000 residents having left the city in a caravan of buses twelve miles long. Six days later, the radius of the evacuation zone was expanded to eighteen miles after tests revealed dangerously high levels of radiation. Then, in early June, twenty thousand more people were evacuated from "hot spots" discovered outside the zone. All totaled, 135,000 people were removed from 179 villages and towns.

Generally speaking, the evacuation was handled efficiently and well. Keeping people in the area immediately after the explosion was well-advised, since the radiation plume blew quite high and most initial fallout settled away from the town. Also, no one wanted 49,000 people standing on street corners waiting for buses. They were safer indoors and, to that extent, the Soviets were fortunate the explosion occurred in the middle of the night when most residents were asleep in concrete dwellings. Prior to evacuation, the roads were sprayed with polymer resin so radioactive dust wouldn't rise as the buses went by. Still, there were problems.

Many evacuees refused to leave their farm animals behind, and military trucks were required to transport nineteen thousand cattle and other livestock. Some residents simply wouldn't abandon their homes, necessitating forced removal, and others hid from their rescuers. One eighty-five-year-old woman eluded authorities and remained in Pripyat for over a month until she was found.

Coping with the demands of resettlement was an awesome task. Initially, each evacuee was brought to an official reception center, decontaminated, medically screened, and given a new set of clothes. Old clothing was destroyed, and suddenly, the evacuees were people with nothing. They had no homes, no possessions, only their lives. Hastily constructed camps lacked electricity and running water. People slept on the floor or several to a mattress. Most knew that in all probability they'd never return home. Even today, despite the construction of twelve thousand new houses and two hundred community facilities, some of the evacuees remain in camps awaiting resettlement. Still, none of their hardship was evident from the air as we flew toward Chernobyl. It was all quite peaceful and idyllic below. Another flock of birds flew beneath us. Nature, I realized, knew no artificial bounds. Then, suddenly, our helicopter began to descend.

We landed in a pasture that had been converted to a makeshift military base. Three or four helicopters and a half-dozen tents were nearby. Off to the side, I saw several communications trailers and, near them, a dozen large plastic bags with blue worksuits inside. The clothes, I assumed, had been worn by military personnel in the danger zone and would be decontaminated before being worn again.

Almost immediately, our helicopter was surrounded by soldiers. Shandala and the pilot got out and disappeared into one of the trailers. Five minutes passed. Then ten. Finally, Shandala returned and told me we'd been cleared to proceed, but I couldn't take photographs. He didn't ask for my camera; he just told me not to use it. Then everyone onboard the copter was given a badge to monitor radiation exposure and a dust mask to guard against inhaling radioactive particles, and we lifted off again.

From there, it was a seven-mile journey to the power station. Six weeks after the accident, four helicopters were hovering above it, still dropping boron and sand to bury the reactor core. We moved closer and began to circle in an ever-decreasing radius above the reactor hall. It looked like a burned-out tenement building, nothing more, and I was struck by how much hoopla had come from something so small. Even the damage to the building seemed limited. There was less debris than a tornado or hurricane might cause. We circled for ten minutes, dropping to within a hundred yards of the ground. Finally, the pilot asked if

I wanted to see Pripyat. I said yes, and we flew toward the town.

Then something happened: I got a little scared; not for myself, but for us all. We passed over Pripyat, and at first glance it was a normal city with white high-rise apartment buildings, parks, schools, and geometrically designed streets. But there were no signs of life. The city was deserted, devoid of people. Nothing was moving except for trees swaying in the breeze.

I assume observers flying over Hiroshima and Nagasaki in 1946 readily understood the devastation below. Death and destruction were visible everywhere. But looking down at Pripyat and the Chernobyl power station, the damage seemed relatively small; one part of one building had been destroyed. Yet from that damage had come radiation, a danger that couldn't be heard, seen, or smelled, and that radiation had rendered an area over a thousand square miles in size uninhabitable.

This is it, I realized. This is what we've been afraid of for all these years. Splitting the atom could be the end of us all.

We circled Pripyat several times. Then the pilot asked if I'd had enough. I told him yes, and he turned the helicopter back toward Kiev to take us home.

CHAPTER
11

Returning to Moscow from Chernobyl and Kiev, I learned that the disaster death toll had risen to twenty-six. Many of the victims had been my patients, and I felt a special sense of responsibility and loss—feelings heightened by a memorial program I watched on Soviet television. It was a two-hour special, showing each of the victims with family and friends. Knowing patients as their physician, I sometimes lose sight of the fact that they've had lives before becoming ill. It's a defense mechanism which, in some way, lessens the impact of their death upon me. But here, on television, I saw my patients alive and well, as they were before Chernobyl.

As the week progressed, I continued to spend considerable time at Hospital Number 6 with Guskova and Baranov. However, my interchange with the Soviet medical establishment had begun to reach beyond Chernobyl. I was invited to lecture on bone marrow transplantation at the All Union Oncology Center, and gave a seminar on leukemia treatment at the Institute for Hematology and Blood Transfusion.

Overall, the general level of health care in the Soviet Union isn't equal to ours. Their medical facilities are older and, in most instances, only limited modern technology is available. Virtually every American hospital has a CAT-scanner, electronic blood-cell counter, and other equipment superior to the average Moscow hospital. Drugs, equipment, and supplies we take for granted aren't even manufactured in the Soviet Union. Significantly, the Soviets invest 3 percent of their gross national product in medical care. In the United States, with a far larger base to begin with, the figure is 11 percent.

Soviet medical personnel are quite good. Intellectually, their best are equal to ours, although I question how deep their ranks are. Still, intellectual capacity isn't the only prerequisite for a good physician. In the United States, we encourage the physician-investigator—someone who treats patients and also works in the laboratory. Soviet doctors do one or the other, not both, and as a result, performance suffers.

Perhaps more important, Soviet physicians are at a disadvantage when it comes to access to information. I'm information-deluged. I get a dozen international telephone calls each day. My computer constantly delivers data, and I read fifteen or twenty medical publications and journals per week. Also, there's a new kind of medical science built on face-to-face contact, not journals, and the Soviets don't participate in that process. Except for rare instances, their physicians are absent from medical conferences in the West. Maybe it's the cost of attending. Perhaps there's fear of an isolated defection. Possibly, the Soviet medical establishment considers itself the center of the world, much like the old Chinese emperors. Whatever the reason, Soviet doctors are disadvantaged in terms of information. Science marches on, and anyone who waits for data to appear in a medical journal will be six months to a year behind.

Still, when the Soviets want to do something well— be it space exploration or microsurgery—they'll spend the dollars to do it right. The All Union Cardiological Institute is superbly staffed, with equipment similar to modern American hospitals. And in several instances, the Soviets may actually be ahead of us medically.

One of the most extraordinary things I saw in Moscow was a hospital "assembly line" at the Moscow Research Institute of Eye Microsurgery. Dr. Svyatoslav Fyodorov, the institute's director, is generally credited with having developed an operation known as radial keratotomy—corneal surgery as an alternative to corrective lenses for people with myopia. In the United States, it's a controversial procedure, but not so in the Soviet Union, where Fyodorov is a medical celebrity and has performed more of these operations than anyone in the world.

After my return to Moscow from Kiev, the Soviets took me to the Eye Institute, where eighty radial keratotomies are performed each day. Most of the patients were men, who went into a dressing room, changed into hospital gowns, and were examined by a nurse one last time. Then they were taken to a conveyer belt with plastic walls separating a series of beds which were literally on tracks. The first patient lay down on the first bed and was covered by sheets except for his eye. Then the operation began.

There were five surgeons on the assembly line. Each one had a specific task to perform, and when it was done, he'd push down an armrest, thereby turning a light on the opposite wall from red to green. At the first station, the patient's eye was anesthetized. One eye was all that was operated on at a time. Then, when the light was green, the plastic wall opened and the bed moved down the track to the next station. The second physician mapped out the spot where an incision should be made. Physician number three made the incision. Number four sutured the cut. Number five bandaged it. As each doctor completed his task, he released the armrest, and when all the lights were green, the beds moved on. If something went wrong, if there was a complication, the patient was taken off the assembly line to a room where another surgeon spent extra time with him. Every few hours, the physicians changed places. At the end of the assembly line, the patients got out of bed, and a nurse brought them to a waiting room where they were given tea and cookies. Then, after thirty minutes, they put their street clothes on and went home. Ten days later, they came back for the other eye to be operated on.

It worked! Eighty operations were performed in the time it would take an American hospital to do four or five. Each of the surgeons was highly competent and able to deal with contingencies as they arose. I guess the more basic question is whether radial keratotomies should be performed at all. Can we afford a one-in-a-thousand complication rate that might cause permanent damage to someone's eye? I don't know the answer, but I'm terribly nearsighted. I wear contact lenses, and if I ever need eye surgery, I'll go to Fyodorov before I go to anyone else in the world.

The rest of my stay in Moscow was a blur of activity. One evening toward the end of the first week in June, I went to the American Ambassador's residence for dinner. Ambassador Hartman and I had met briefly in George Shultz's office in Washington, but this was a longer encounter. It was a small dinner attended only by Hartman, his wife, and one other guest, and afterward Hartman drove me back to the Sovietskaya in his Mercedes. We got in the car and I said, "There are several important things I didn't think it was appropriate to discuss in front of the guests at dinner." Hartman's response made it clear that, in his view at least, my fear of eavesdropping wasn't paranoia. "If they're important," he answered, "don't tell them to me in this car."

That same week, I met the Soviet poet Yevgeny Yevtushenko. Rick Jacobs was friendly with him and had suggested I make a courtesy call. Yevtushenko lived in Peridelkino, a writer's retreat about thirty minutes from Moscow, in a dacha that was well-appointed but in total disarray. Soon after I arrived, he proposed that we visit

Boris Pasternak's grave. This was before Pasternak's "re-habilitation" in Communist Party circles. Regardless, I agreed, and we drove to a small cemetery nearby. There was a modest marker with a portrait of Pasternak carved in stone, and a modestly dressed old man tending the grave. He'd been Pasternak's housekeeper and, since the writer's death, had come to the cemetery every day to recite Pasternak's works and keep the flowers in bloom.

After visiting Yevtushenko, I returned to Moscow, where one last task remained. Toward the end of my first trip, I'd suggested to Victor that the long-term medical consequences of Chernobyl were of major concern, and that it would be in everyone's interest to put a mechanism in place to deal with the issue. More specifically, I proposed long-term evaluation and medical care for the 135,000 residents evacuated from the danger zone. Victor relayed my idea to the appropriate Soviet authorities and their response was positive, but that left open the issue of which organizations would be involved.

I was a private citizen. Obviously, I couldn't enter into a comprehensive program as a representative of the United States government, and the Soviets didn't want a medical-evaluation agreement with our government anyway. On the other hand, though, it wouldn't mean much if I signed only as Robert Gale. The question had been unresolved when I left Moscow after my first visit, although I'd contacted the respective heads of the National Institute of Health and National Academy of Sciences with an eye toward resolving the matter. Both organizations expressed interest, and at the start of my second trip to Moscow, I'd spoken extensively with Andrei Vorobiev of the Central Institute for Advanced Medical Studies. Finally, we agreed on the text of a document which called for a joint study of the medical consequences of Chernobyl and stated that each of us would enlist the aid of appropriate organizations in our respective countries in furtherance of our effort. I signed as a UCLA faculty member and chairman of the Advisory Committee to the International Bone Marrow Transplant Registry. Vorobiev signed as a department head of his institute and a member of the Soviet Academy of Medical Sciences.

The agreement was announced at a June 7th press conference just prior to my leaving the Soviet Union for the second time. As before, we followed a question-and-answer format after brief opening statements. Then, to celebrate, I hosted a dinner for Baranov, Guskova, Victor, Vorobiev, and a few others who'd been at the heart of the Chernobyl relief effort.

One of the complications in going out for dinner in the Soviet Union is that some restaurants take only roubles and others only foreign currency. For our dinner, I chose the National Hotel, where Mikhail Bruk had entertained us several weeks earlier. Victor was a bit nervous. He kept asking me whether the hotel restaurant accepted foreign currency, and shouldn't we make an effort to find out before dinner. Finally, with assurance, I told him, "Victor, you have it backwards. First, we eat dinner. Then the restaurant has to take my currency. They can't throw me in jail, because I'm wanted at the hospital."

Dinner was a farewell of sorts, since I'd be returning to Los Angeles the following morning. We sat down at sunset, with all of Red Square framed through a picture window, and each of the Soviets presented me with a gift. Baranov gave me an antique samovar, Guskova a two-volume collection of Soviet literature. There were art books, small sculptures, all quite nice.

We had a traditional Russian dinner, consisting primarily of vodka and zakuski. In the United States, hors d'oeuvres tell us we're getting ready to eat. But with the best Russian meals, the zakuskies are it. Usually, there are ten or fifteen items—sturgeon, caviar, cucumbers, stuffed eggs. Then the main course deteriorates to something like boiled beef.

We drank a lot of vodka, and there were many toasts. Finally, at the end of the meal, I presented my gifts. Back in the United States after my first Soviet journey, I'd commissioned special T-shirts. The manufacturer had used English letters, since T-shirt makers in Southern California don't keep the Cyrillic alphabet in stock. And what the T-shirts said was, "CCCP-USA International Bone Marrow Transplant Team."

It was a uniquely American gift. People in Moscow don't wear T-shirts. And I was determined that they go only to key participants. In Los Angeles, I'd give one each to Dick Champlin, Paul Terasaki, Armand Hammer, and Rick Jacobs. I'd kept one for myself and mailed another to Yair Reisner in Israel. Now I went around the table after dinner, handing everyone a white T-shirt with "CCCP" in red letters and "USA" in red, white, and blue. I sort of wondered what they'd do with them, and Baranov was the first to act. Not bothering to take off his suit jacket, he pulled the T-shirt on over his head, and a new fashion was created among the Soviet elite.

On Sunday, June 8th, I flew back to Los Angeles. There, for the rest of the month, I was "attending physician" at the Bone Marrow Transplant Unit, which meant most of my time was spent in the ward at UCLA. Since I was planning a third trip to Moscow in mid-July, I had my usual shopping list of items to buy—more Polaroid film for Baranov, croissants for Fetisov's daughter. I also formulated concrete plans to implement the agreement Vorobiev and I had signed regarding the long-term evaluation and medical treatment for people from the evacuation zone.

At one point, I used my newfound international influence (such as it was) to expedite a legal proceeding for my wife, Tamar. Several years earlier, at my urging, she'd applied for United States citizenship. She was quite happy as an Israeli citizen, but on occasion I visit countries that don't honor Israeli passports, which made it difficult for her to come with me. Also, each time we returned to the United States, Tamar had to wait in a long immigration line.

Tamar's first application for American citizenship was lost by the Immigration and Naturalization Service. There was a two-year delay in processing her second application, after which the service advised us that the application was deficient because the accompanying photo was now more than the legally mandated one year old. Finally, after I got back to Los Angeles from my second Soviet journey, I telephoned the State Department and told them, "Look, I'm going back to Moscow in July. I want my wife with me, but there's a problem because she has an Israeli passport. Please, do something." Several days later, I got a telephone call, telling me Tamar's application had been approved, and she took the oath of citizenship the following Friday. That, by chance, fit nicely with another event in our lives—the Statue of Liberty Centennial celebration.

Some time earlier, I'd decided we should bring the children to New York for the relighting celebration. It wasn't important to attend any of the special events, but I wanted the kids to be in the city to experience the atmosphere of the occasion. Finding a hotel was horrendous, but we did it. We were regular tourists and, on the afternoon of July 3rd, took a ferry from Manhattan to Staten Island and back again. Then I called the hotel to check for messages and learned that CBS had left two tickets to the evening relighting ceremony for me. That began a mad dash back to the hotel to change clothes, find a babysitter, pick up the tickets, and get to a pier in Brooklyn where the boat to Governor's Island was departing at six o'clock. Tamar and I made it, although not by much, and overall the relighting ceremony was quite moving.

Part of it was less than wonderful. It's one thing to turn the closing ceremony at the Olympic Games into eighty-four pianists playing Gershwin, but some things are too serious to trivialize as entertainment, and liberty is one of them. Sacred values are ill-served by two hundred Elvis Presley impersonators gyrating around a football field, which is what Liberty Weekend ultimately degenerated into. American society and the men and women who've died to defend it shouldn't be reduced to the level of a musical.

Still, most of the relighting ceremony was very touching. I kept thinking back to the statue of "Mother Russia" in Kiev, and the relatively small distance between the Soviets and ourselves. I focused on how lucky I was to have been born in the United States and the fact that it was largely by chance that my great-grandfather had emigrated from Byelorussia to New York.

One thing in particular struck me about the evening. My tickets had come courtesy of CBS, but sitting next to Tamar and myself, there was a couple from Kansas. Neither of them had any special political connection or prominence. So far as I knew, they weren't wealthy. They were ordinary Americans who'd decided a year earlier that they wanted to attend the relighting ceremony, bought their tickets, and now they were sitting within fifty yards of the President of the United States as he relit the torch of the Statue of Liberty.

That couldn't happen in the Soviet Union. For all their talk of a classless society, if you're not well-connected, there's no way you can get into most clubs in Moscow. To eat in the right restaurants, to attend certain premium events, you need a position in Soviet society. Even Baranov, who's a talented physician and loyal party member, would find many gatherings off-limits.

By contrast, in the United States, people make reservations for dinner at the restaurant of their choice, and if they have the money to pay for it, they're seated. They can be nitwits, they can be members of any religion or race, and still they'll be admitted. A small number of private clubs are inaccessible, but overall, money is a leveling factor. That's not to ignore the economic injustice that plagues our society. Often, money is a very high barrier, and wealth should be distributed more equitably than it is in our country. But at least Americans don't have to be well-connected, we don't have to be government officials, and we don't have to be related to somebody to attend a national ceremony. Everyone has the right to buy a ticket. That doesn't happen in the Soviet Union, where they have an established order and it stays that way. Soviet citizens can't create their own world or succeed outside the system in their country.

On Thursday, July 17th, I flew back to Moscow. Hammer was going to attend the reopening of his art exhibit in Novosibersk and included me in his entourage.

Victor met us at Sheremetyevo Airport and drove me to the Sovietskaya. Unlike most occasions when he seemed tireless and imperturbable, this time he was borderline-crazed as the result of a domestic problem. During my absence, he and his family had moved to a new apartment, but couldn't unpack because the floors had to be scraped, sanded, and refinished. In the United States, he simply would have gone to a hardware store, rented a sander, and done it himself. But no machines were available in Moscow, so he called the official state floor-scrapers union (or whatever it's named), which said it would be several months before they could do the job.

Victor didn't want to wait. He and his family were living out of boxes and wanted to get settled, so he began looking for freelance help—someone who worked for the official union but would moonlight at night. Finally, he found someone who agreed to the task, but only on the condition that Victor be there to aid in the effort. Thus, Victor, his wife, and daughter were mired in cartons, constantly inhaling wood shavings, and performing manual labor until midnight each night.

On Saturday, July 19th, I flew with Hammer on Oxy One to Novosibersk. This was the first exhibition of foreign art to open in Siberia, and most of the city turned out to greet us. I'd expected to be recognized in Kiev but not in Siberia, and was struck by how many people knew

146

my face and name. It made me realize how deeply Chernobyl had scared the Soviet people.

Hammer gave a brief speech to open the exhibition. Then we toured the city, ate dinner, and flew back to Moscow. The following day, Sunday, I had a morning interview with a reporter from *Izvestia*, and the rest of the day was "vacation"—my first in the Soviet Union. After weighing my options, I took some apples and cheese, and walked over to Red Square to sit in the sun.

What do Russians do on a Sunday afternoon? They talk; they laugh; they eat ice cream; their children run around. Red Square was jammed. Most people were wearing casual clothes, but without the extremes in dress you'd find in a comparable American setting.

Red Square was mostly concrete and stone, more like the courtyard at St. Peter's in the Vatican than an urban park. Several people noticed me and smiled, made eye contact, or asked for an autograph. Near the perimeter of the square, I saw a machine where drinks were for sale. It was very odd. Three choices were offered—cider, sparkling water, and some kind of soda—but there was only one glass in the machine. It didn't have disposable cups. People stood on line, put their money in, pushed a button, and a little bit of water came down to rinse out the glass. Then they put the glass back, pushed another button, waited for the glass to fill up, drank their drink, and left the glass for the next person.

I sat on a wall maybe four feet high and watched people go by. It was my first time alone in the Soviet Union without a translator, driver, or someone else monitoring me. Nearby, several dozen people were standing in line for an ice cream vendor. It was a nice afternoon, an opportunity to relax, think, and be alone.

That evening, I went back to Hammer's apartment for a cocktail party before attending the closing ceremonies at Moscow Stadium for Ted Turner's "Goodwill Games." Monday morning, I was at the Ministry of Health to discuss implementation of the agreement I'd signed with

Vorobiev several weeks earlier. Then I went to Hospital Number 6, where Baranov met me. "Varsinian died this morning," he said.

Of all the blows suffered in the aftermath of Chernobyl, none hurt more than Varsinian's death. Most of the patients were cured or died within two months of the accident, but Varsinian had lingered in a never-never land between despair and hope. At times, his prognosis was decidedly optimistic. Initial indications were that his bone marrow transplant had successfully engrafted, and when I'd left Moscow after my second trip, he was doing well. In fact, a photograph of Varsinian, alive and smiling in the sterile unit, appeared on the cover of *Life* magazine. Then, in early July, he'd taken a turn for the worse. In Los Angeles, I'd received a telex from Baranov, saying Varsinian's blood count had fallen precipitously, asking what steps I would recommend. There was a paradox in that Baranov thought Varsinian had graft-versus-host disease, which is generally associated with a stable blood count, but inconsistencies of that sort happen sometimes; we don't know why. All I could do was telex back that we've seen this before and this is how we treat it, although we don't know exactly what's going on.

It wasn't enough. Varsinian, the patient we'd gotten to know best, one all of us were immensely fond of, died. "I'm going to the autopsy," Baranov told me.

In the United States, physicians rarely attend their patients' autopsies. Only 10 percent of deceased persons are autopsied to begin with. And beyond that, doctors don't like to face their failures. It's very upsetting to see someone you've treated and cared for lying on a slab of stone. Hours before, they might have been very sick but they were alive, and now they're being sliced open. Their life is over. A person can be alive and sick, alive and dying, comatose and on a respirator, but they're still alive. And then, in a single moment, that absolute line between life and death is crossed, and they're gone forever.

As painful as the experience is, my colleagues at UCLA

and I attend the autopsies of every patient who dies. It's essential. We have to understand what went wrong, and many mysteries are unsolvable while a patient is alive. Why was he breathing so fast? Was there fluid in his lungs? Bleeding in the brain? After death we can put tissue and cells under microscopes to search for answers. It's torture, but we learn.

In Varsinian's case, death was caused by multiple organ failure, but was that the result of graft-versus-host disease, a disseminated viral infection, or both? The answer was enormously important insofar as determining how future radiation victims should be treated.

I hadn't attended the autopsy of any previous Chernobyl victim. This time, I told Baranov I wanted to be there. We drove to a building not far from Hospital Number 6, an old morgue-like structure, although I'm not sure it was a morgue. There, we put on surgical outfits—boots, white coats, masks, hats, gloves—and entered the autopsy theater.

Varsinian was lying on a stone table. Twenty people were in the room, which is a lot for an autopsy, and a small film crew was standing by with lights and a camera. The room itself was fairly large, twenty by thirty-five feet in size. There were three autopsy tables, each one with a drain and stainless-steel rim, and quite a few cabinets along the walls.

The autopsy began as soon as Baranov and I arrived. The prosector made a parasternal incision in the torso, and one by one major organs were removed. The table had a faucet and attached hose at one end. Each organ was washed and weighed after removal, then examined for abnormalities. Pieces of tissue were put in jars. A pathologist wrote down every comment and observation that each of us made. Always, there was the sound of running water.

Varsinian wanted very much to live, and all of us at Hospital Number 6 had done everything in our power to save him. Most of the patients we treated are well now, but we don't see them. The ones who stay with us are the ones who died.

As a doctor, my failures haunt me. There's a biological reason why some patients live and others expire. It's not just chance, and with each death I find myself asking, what could I have done differently to save this person? Each failure erases ten successes because, having saved one patient, I hope always to succeed. I'd decided that I wasn't going to allow Varsinian to die. And in the end, nothing we did could keep him from dying.

CHAPTER

Varsinian's death discouraged all of us at the hospital, but my spirits lifted when Tamar and the children arrived in Moscow.

It had been my idea to bring the children. Tal and Shir were nine and seven years old respectively. Elan was a few months shy of three. I don't like being away from them too often, and thought it would be a good experience for them to see the Soviet Union. Meanwhile, the Soviets were delighted, since their coming signaled to the world that Moscow was safe in the aftermath of Chernobyl. They even gave me a bigger suite at the Sovietskaya to accommodate everyone, and arranged visits to the zoo, the circus, and other attractions for the children.

Guskova in particular got along with the kids. She's a strict disciplinarian, the sort of person who will play with a child for an "appropriate" length of time, and then say "that's enough" in a very authoritarian manner. But there's a reservoir of warmth beneath her rather gruff exterior, and at her urging, Tal became pen pals with her grandson, Sasha.

I liked Guskova. She was a proud woman; proud of herself and proud of being Russian. Once, she told me that the copper used to make the Statue of Liberty had been mined in the Ural Mountains and brought to Paris for casting. The gifts she gave were always Russian—a book of poems by Lermontov, a rock from the tundra. On the day she and Hammer met at Hospital Number 6, she gave him a plain wooden bowl with salt in it—a traditional Russian offering. To her, it was beautiful, and she was proud to give it to one of the richest men in the world.

On Wednesday, July 23rd, Tamar, the children, and I flew to Kiev with Hammer. Several of the Soviets went with us and were astounded by the open bar, buffet lunch, and VCRs on the Occidental plane. They'd never seen anything like it in the air before. It was on this flight that Elan developed the habit of kicking Hammer in the shins when he wanted attention. Hammer was gracious about the whole thing, and said Elan would do very well in life because he knew how to make his desires known, but to this day he talks about the fact that Elan kicked him in the shins. Actually, I'm not sure what it was Elan wanted. I could ask, but since he was two years old at the time, he probably doesn't remember anymore.

Shortly before noon, we landed in Kiev, and a police motorcade led us to the heart of town. That afternoon, Hammer and I visited Hospital Number 14 and the medical museum. Then we returned to the airport for a helicopter flight over Chernobyl. I'd already been there, but Hammer wanted to see it and asked me to come along. Victor and Mikhail Shandala also went for a second time, and we were joined by Ivan Nikitin, Anatoly Romanyenko, and John Bryson, a freelance photographer affiliated with *Life* magazine.

The copter was larger than the one I'd previously flown in, and also faster. The trip over and back took a little more than an hour. Victor's reaction to seeing Chernobyl was great sadness, as it had been during our first flight in early June. Then Hammer flew back to Moscow, and I stayed in Kiev with Tamar and the children.

The next morning, Thursday, July 24th, I visited the children's ward at the Institute of Pediatrics, Obstetrics and Gynecology. Tal gave a violin concert for some of the patients, and Shir accompanied her by singing Hebrew songs with Elan's assistance. Then we distributed gifts to the children—Los Angeles Dodgers baseball caps, children's toys—things that were distinctly American. Late in the day, Victor took me to a soccer game between Dynamo Kiev and Dnieper. Dynamo Kiev won 2 to 1, and I'm not sure whether I was invited for some complex political reason or simply because Victor wanted to see a soccer game.

The following days were a mix of business and tourism. I went back to Babi Yar with Tamar and the children, then to "Mother Russia" and the Tomb of the Unknown Soldier. Meanwhile, I was trying to find out what effect, if any, Chernobyl had had on the water and food supply around Kiev and anything else I could learn. On Sunday, July 27th, we returned to Moscow, and the following morning I visited the All Union Oncology Center to discuss plans for an upcoming satellite conference on cancer treatment. Then I went to Hospital Number 6, where Baranov told me that a plant worker named Novik had died.

Novik was the last of the transplant patients to die. By this point, all the others were on the road to recovery or had preceded him to the grave, so now, even more than before, we began focusing our attention on medical liaisons and long-term studies. On Tuesday, July 29th, I presented seminars at the Institute for Advanced Medical Studies and the Institute for Hematology and Blood Transfusion. I also opened negotiations for Soviet participation in the International Bone Marrow Transplant Registry, which eventually came to fruition. Significantly, the Soviets were influenced by the fact that they wouldn't have to pay to join, and were pleased to learn that the registry actually paid one hundred dollars for each case reported.

My last set of negotiations involved the June 7th agreement I'd signed with Vorobiev. Basically, it called for the long-term evaluation and care of people from the evacuation zone, but there were no teeth in it. Nothing was concrete, and the question arose, where do we go from here? What sort of guarantees accompany the study? How will it be implemented?

The Soviets' position was, we have an agreement saying we'll work together; that's enough. We don't need another agreement; nothing has to be spelled out further in writing; these things will happen as they happen. I wanted more and, at a series of meetings, pressed hard for specifics. All totaled, I conferred with Vorobiev, Nikitin, Schepin, and Minister of Health Sergei Burenkov a half-dozen times, but they held firm. "We have an agreement," they told me. "That's enough." Their response was disappointing, but there was nothing I could do to change their minds.

On Thursday, July 31st, my third Soviet odyssey came to an end and I flew to Paris with the children and Tamar. From there, we took a flight to Israel and stayed in Jerusalem with Tamar's parents for five days. It was an opportunity for me to visit old acquaintances at the Weizmann Institute, and I saw Yair Reisner for the first time since May. The Israeli press had paid considerable attention to his exploits in Moscow and, returning home, he'd been a hero, as he put it, "for about a day."

I have very strong feelings about Israel. They're not lifelong; many of them came to the fore only after I married Tamar. But to me it's quite simple: Israel has to exist. There's nothing I'd gladly die for and very few things for which I'd be willing to risk my life, but Israel is one of them. It's a nation whose existence and people are constantly threatened, and none of us dare allow the Holocaust to happen again.

It's a matter of fundamental right and wrong. I don't live in Israel; the United States is my country and my home. But if needed, I'd get on a plane and fight for Israel's survival tomorrow. Our three children, all born in the United States, have dual citizenship, which means they'll serve in the Israeli Army. Tamar and I thought about it at length before we made the decision. When Tal, Shir, and Elan are eighteen, they can do what they choose. And if they decide it's of paramount importance to be personally safe, they can renounce their dual citizenship and be free of Israeli military obligations.

Don't get me wrong, I know there are problems with Israeli society. But I traveled through Lebanon just after the invasion of Sidon. Beirut wasn't a city anymore. It had been turned into a military camp by the Palestinian Liberation Organization. I lived in Israel for a year with my wife and daughters. When my children's school arranged a class trip, one of the parents had to go along with a loaded submachine gun to guard against terrorism. Every morning before school, other parents and I took turns surveying the school grounds for explosives. Two of my children's playmates and the mother of a third were killed in a terrorist assault. These are things Americans don't have to live with, and at some point they become more than philosophical considerations. So my view is, people can sit back at cocktail parties and talk about the West Bank, but that's very different from walking around your child's school at six in the morning looking for bombs. I also believe that no one is more concerned for the Palestinians' well-being than the Israeli people, for reasons of self-preservation if nothing more.

After our trip to Israel, I spent two weeks in Los Angeles before going to Copenhagen. There, I attended a conference on human rights devoted to the issue of whether or not people have a right to receive humanitarian medical aid. In other words, do nations have an obligation to offer aid in time of disaster, and do governments have the ob-

155

ligation to accept such aid on behalf of their people? Obviously, I'd been invited to participate because of Chernobyl, but that was just one aspect of the debate, and there was ample legal precedent for our deliberations. Several countries, including the United States and the Soviet Union, have treaties calling for assistance if an astronaut or cosmonaut from one country lands on another's soil. And, of course, ships at sea in distress are aided by other vessels.

The conference concluded that the right to receive humanitarian aid such as food and medical supplies is a basic human right. A proposed amendment to the United Nations Bill of Rights incorporating that position was drafted and later presented by the Danish delegation to the United Nations for General Assembly approval. Then I returned to New York, where the Soviet exhibition of Impressionist art was opening at the Metropolitan Museum after showings in Washington and Los Angeles. An Occidental Petroleum limousine picked me up at Kennedy Airport and brought me to the Plaza Hotel, after which I attended a formal dinner in the museum courtyard as a guest of Dr. Hammer. Then, the following morning, I went jogging up Fifth Avenue in Manhattan, reflecting on the fact that at times it seemed I did nothing but follow the Hammer and Soviet art exhibits around the world. The sun was rising as I ran, and to the side I saw twenty or thirty homeless people asleep on benches beside shopping bags piled high with rags. Many of those people, I knew, were mentally ill and should have been hospitalized. Others were on the street because of unfortunate and frightening circumstances in their lives. But whatever the reason, it was disturbing to me that in a society as rich as ours, people sleep on the streets. There's no excuse for it; this isn't Bangladesh. The Soviets take film footage of Americans huddled over grates in Washington, D.C., or sleeping in doorways in New York, and they put it on television to attack us around the world. My instinct is to defend the United States at all times, but some things are very hard to defend and the plight of our homeless is one of them.

The next major Chernobyl-related event in my life occurred from August 25th through the 29th, when the International Atomic Energy Agency met in Vienna to review the disaster. Every nation with an interest in nuclear power was represented. I was there as part of the American delegation.

One of the leaders of the Soviet delegation was Leonid Ilyin, a radiation specialist and vice president of the Soviet Academy of Medical Sciences. At the start of the conference, he presented a comprehensive 430-page report explaining how and why the accident happened, and virtually everyone in attendance was impressed with the amount of data offered.

Previously, much of the world had been critical of the Soviets for their delay and subsequent failure to disclose adequate information about Chernobyl. The first indication that something had gone wrong at the power station came two days after the accident, when abnormally high levels of radiation were detected in Sweden. An analysis of radioactive samples revealed the presence of cesium-134, a fission product found only in nuclear reactors. Prevailing winds suggested a source in the Soviet Union. That afternoon, April 28th, the Swedish Ambassador in Moscow had telephoned the Soviet State Committee for the Use of Atomic Energy and was told, "We have no information we can furnish to you." Later in the day, in response to further inquiries, a representative of the Soviet Foreign Ministry told Swedish officials, "We have heard nothing."

In truth, Soviet leaders had known enough about the disaster two days earlier to treat it as a major crisis and establish a high-level commission to direct recovery operations. But it wasn't until the evening of April 28th, sixty-seven hours after the accident, that a fourteen-second announcement broadcast on Radio Moscow declared,

"An accident has occurred at the Chernobyl nuclear power plant. One of the atomic reactors has been damaged. Measures are being taken to liquidate the consequences of the accident. Those affected are being given aid, and a government commission of inquiry has been created."

Thereafter, information was released slowly and in stages. On April 29th, the sixth item on a Soviet television news broadcast stated that two people were killed in the accident, part of the reactor building was destroyed, and residents from Pripyat and three neighboring towns had been evacuated. One day later, *Pravda* and *Izvestia* ran small articles about the accident on inside pages, but nothing more of significance was said until May 6th, when an extensive report detailing the heroics of Soviet firefighters appeared in *Pravda*. Further revelations followed, but most were aimed at minimizing the tragedy and placing it in a context consistent with Soviet political ideology. No film or photographs of the dramatic evacuation were released.

In contrast to all this, the Soviet presentation at Vienna was remarkably thorough and candid. Gross operating errors at Chernobyl were acknowledged, as were a lesser number of reactor design flaws. Soviet officials revealed that it had taken twelve days to extinguish the fire in the reactor core, and admitted that quite a few new fires had broken out in the reactor building since then. Equally compelling were accounts of Soviet efforts to halt radiation leaks, which continued through early May.

From the start, Soviet authorities faced an unprecedented task—extinguishing a major fire inside a nuclear reactor while vast amounts of radiation were escaping and three more reactors stood nearby. Chernobyl Unit Number 3 was taken out of service three and a half hours after the accident. Units 1 and 2 were shut down at 1:13 the following morning. Meanwhile, at dawn on the day of the explosion, Soviet scientists conducted reconnaissance flights over the reactor building to evaluate the damage and formulate a strategy for handling the crisis.

Their conclusion was that the fire could only be extinguished from the air, and over the next two weeks, five thousand tons of lead, clay, boron, dolomite, and sand were dropped by helicopter to cover the exposed core, ensure that the rubble remained subcritical, and smother the blaze.

However, resolution of one problem led to another. After the explosion, a huge basin beneath the reactor filled with water from the reactor cooling system. And once five thousand tons of sand were dropped and covered the core, it was increasingly likely that the reactor would collapse downward into the basin and interact with the water to produce radioactive vapors. In a worst-case scenario, that would lead to a meltdown through the basin floor.

Thus, three engineers were called upon to drain water out of the basin, and four hundred miners working around the clock over the next ten days tunneled 160 yards to an area beneath the reactor, where they installed a thirty-yard-square concrete slab and nitrogen cooling system. Subsequently, the entire reactor was entombed in a 300,000-ton concrete sarcophagus to prevent the further escape of radioactive particles. All of this was accomplished by the end of June, but even then the Chernobyl cleanup had just begun.

In essence, the Soviets had to decontaminate the entire evacuation zone—a Herculean job requiring the largest peacetime mobilization in history. Tens of thousands of workers were involved. Radioactive topsoil was skimmed by bulldozers from surface areas, carted away, and stored in drums. Sixty thousand buildings were scrubbed down. Trees were sprayed with decontaminants and fallen leaves buried underground. Clouds were seeded before they reached the danger zone to prevent the spread of radiation through the runoff of rain. Concrete barriers were built to prevent contaminated water from seeping into the Pripyat and Dnieper rivers. All totaled, the cost of the accident to the Soviet people in lost electric power, resettlement, and clean-up expenses has been estimated at three to six billion dollars. Neighboring countries claimed a half-billion dollars in spoiled crops and related damage. In Scandinavia, fifteen hundred reindeer which fed on radioactive grass were slaughtered and buried in a mass grave beneath a marker reading "Hazardous radioactive materials underground."

All of this weighed heavily on the Soviets in Vienna. Then, too, they had to acknowledge the medical consequences of Chernobyl. The medical material was presented in part by Guskova, and she made several statements which in my view were not entirely accurate. Guskova is an excellent physician. She's also a reasonable woman, so I have to believe there were reasons other than scientific for her report. More specifically, I think she was guided in her relatively negative evaluation of bone marrow transplants by political rather than medical considerations.

My own professional career has focused on diseases rather than treatments and technologies. Bone marrow transplantation is simply a technology. Some people think you can best cure cancer with chemotherapy. Others swear that transplantation is the answer. I don't care about these rivalries. I'm concerned with the disease and have no vested interest in any particular cure. If someone discovered that we could cure leukemia by throwing patients into swimming pools, I'd be at poolside tomorrow.

I say this because I had no particular ax to grind at Vienna, but Guskova seemed under considerable pressure to downgrade American assistance to the victims of Chernobyl. And, of course, one way to do that was to cast doubt on the bone marrow transplants. In reality, the transplants weren't our major contribution. The media seized on them because they were an extreme intervention, but other contributions were equal if not greater in value. Still, the Soviets didn't want it to appear as though Americans had come to rescue them. And in addition to being rather cool toward me at Vienna, Guskova voiced conclusions about transplants which overlooked the fact that they'd been given only to patients who had no chance of surviving without them.

The Vienna conference marked the close of an era. For the first time, the Soviets had elaborated on the causes and effects of Chernobyl, and acute medical events were at an end. Still, thirty patients remained in Hospital Number 6, and six weeks later, in early October, I was in Moscow again.

There were two reasons for my fourth journey. One was to maintain contact with Soviet medical personnel, who could help in pursuing the long-term evaluation and treatment of evacuation zone residents. And second, Hammer's ubiquitous art exhibit was opening in Kiev, and he wanted me there for the show.

I spent a day in Moscow with Baranov, who was now consulting me with regard to patients having nothing to do with Chernobyl. By this time, there was near-absolute trust between us, and the only flaw I saw in his character was an unhealthy inclination to chain-smoke compulsively. Indeed, my most lasting visual image of Baranov is of his creating ashtrays out of virtually anything at hand. Lighting a cigarette, he'd cast about for something functional—a cup, an empty can. Often, he'd take a piece of paper, fold it in complex fashion so it became an ashtray, and, when finished smoking, simply crumple it up and throw it away.

My most gratifying moments with Baranov were when we talked openly with each other; when he solicited my advice, and I his. Once, he asked if I thought the Chernobyl victims would have fared better if they'd been treated in American hospitals. And I answered honestly: "Probably not. Maybe two or three more would have lived; that's all." Another time, we had a conversation that still haunts me. Baranov asked what I thought would happen if either the United States or Soviet Union unilaterally destroyed all of its nuclear weapons. I said I didn't know, and he told me, "I think whichever side destroyed the weapons would be attacked by the other."

Probably, the most moving events to occur during my fourth trip involved the circumstance of Jews in the Soviet Union. Closed borders are a hallmark of totalitarian society, and in recent years, Jewish emigration has been severely restricted by Soviet authorities. In the 1970s, an average of twenty-five thousand Jews per year were allowed to leave. However, in the 1980s, Jewish emigration has been drastically reduced, sometimes to as little as one thousand per annum.

The issue of Jewish emigration is one I've been forced to deal with ever since Chernobyl. No one knows how many Jews there are in the Soviet Union. Best estimates range from two to three million. And no one is sure how many of these want to leave. Some say fifteen thousand, which is the number who have formally asked permission to emigrate. Others claim up to four hundred thousand would apply for exit visas if their chances of receiving them were better and there was no fear of government retaliation.

Regardless, when I returned to Los Angeles after my first Soviet journey, virtually every major Jewish organization in the United States contacted me to see if I'd help them. Strategies varied. Some groups felt it was important to work person by person. If we got one individual out, we'd save the world. Others wanted to concentrate on key people like Anatoly Shcharansky and Yelena Bonner. A third viewpoint was, unless the Soviets let tens of thousands of people go, it didn't matter at all. I was also besieged by personal calls: Can you get my mother out? What about my brother? And for everyone who telephoned, their particular relative was the most important person in the world.

Ultimately, I decided it would be wrong to choose one person or one organization over another, and I gave the same treatment to everyone. When someone contacted me, I told the person, "Give me a letter addressed to Mikhail Gorbachev, and I'll personally hand it to a Soviet official who will forward it to him."

In the months that followed, I brought literally hundreds of letters to Moscow. Most were unsealed. By force of habit, few Russians send sealed letters because it's assumed they'll be torn open and inspected by the authorities. I didn't read them but quietly delivered each one, and a handful of Jews were allowed to emigrate as a result of my intervention. It was a privilege to help them, but the only true solution will come when Soviet society changes to the point where it no longer tries to lock people in.

On the night of October 12th, just after the start of my fourth Soviet journey, I had one of the more dramatic experiences of my Chernobyl odyssey. It was Yom Kippur, and Rick Jacobs, who'd accompanied Hammer to Moscow, announced his intention to attend services at the Central Synagogue. Then Tamar, who virtually never goes to temple, decided to go with him to express solidarity with Soviet Jews. That left me wondering what to do. I'm not at all into the formalities of religion. I eat regularly on Yom Kippur, and hadn't been to a High Holy Days service since I was bar mitzvahed at age thirteen. I consider myself religious, but probably no one else would. Clearly, I wouldn't have gone to temple if I'd been in Los Angeles, but the question now was, should I go? I assumed Soviet authorities wouldn't appreciate my going. No one had said anything to me about it, so that was just my view, but I think it was accurate and I wasn't out to antagonize the Soviet government. On the other hand, I thought Tamar's point was worth making; normally I'd go with her and Rick if they went out at night, and not going would have been a concession to Soviet pressure.

So we went together, arriving at sunset. Several policemen outside the synagogue were bewildered by the advent of a large black limousine with three passengers, but they let us in. The temple was an old building, once grand, but a victim of disrepair in recent years. And it was jammed, standing room only, with a third of the congregation under thirty years old. Most of the men wore ill-tailored brown or gray suits; the women, dresses, equally solemn for the occasion. Downstairs, they sat together in thirty rows of battered wooden pews. The balcony was segregated, for women only.

Just inside the front door, a middle-aged man gestured for us to follow him. I don't think he knew who we were. More likely, he was simply taking us to seats because we were strangers. Then someone recognized me. I heard a murmur—"Dr. Gale, Dr. Gale"—and a push of hands carried us to three seats in front of the congregation right by the altar. Once we were seated, another man, this one extremely old, handed us three prayer books written in Hebrew—a rare commodity in the Soviet Union. I speak Hebrew. I learned it because when the children were born, Tamar insisted we speak Hebrew at home so they'd grow up bilingual. If I wanted to eat or talk to my children at dinner, I had to learn. Now, I sensed a tremendous outpouring of warmth from the congregation. They were proud I had come, and even more pleased when they realized during the service that I could speak and read with them. Sitting in the synagogue, I realized what a terrible mistake it would have been not to come, and I'm grateful to Tamar and Rick for challenging me to join them.

One night later, I had a more lonely but equally dramatic encounter. A woman named Rimma Bravve telephoned my room from the hotel lobby and asked to speak with me. I knew her name because her mother had asked me for help. Rimma was thirty-one years old and suffering from advanced ovarian cancer. Her father was dead. Her mother had emigrated from Russia to the United States six years earlier to practice as a pediatrician in Rochester, New York. Rimma had tried to leave Moscow for seven years, but to no avail. "I don't want to come upstairs in the hotel," she told me. "Please, could we meet by the river?"

I agreed, and went to the Moscow River, looking for a woman in a red dress, which is what she said she was wearing. I found her; we talked, and it was very sad. Her cancer was terminal. The Soviets had done what was medically appropriate, but she was beyond help. She was dying and knew it.

"You have to understand," I told her. "Coming to the United States won't change anything medically. Whatever we do can be done for you in the Soviet Union."

"I know. But if I'm going to die, I want to be with my mother."

I told her I had a letter from her mother, which I would forward to the appropriate Soviet authorities. Then I gave her Baranov's name and telephone number in the event that she wanted further treatment. In December, she was granted an exit visa and came to the United States, where she died six months later.

What will happen to Jews in the Soviet Union? I don't know; I wish I did. Many Soviet Jews don't want to leave. They love their country and want to stay there. And, of course, the majority of those who do emigrate come to the United States, not Israel. All I can say is, the Soviet Union is about seventy years old and the Jewish people are six thousand years old. Maybe that's the answer.

My October trip to the Soviet Union ended with typical Armand Hammer drama. On the morning of the 14th, we were scheduled to fly to Kiev on Oxy One for the opening of Hammer's art exhibit. This was an event of particular importance for the Soviets. At our first press conference, five months earlier in Moscow, Hammer had announced that the exhibit would be going to Kiev despite the Chernobyl disaster and that he personally would attend the opening. This endeared him enormously to the people of Kiev, who felt very much under siege at the time. And the planned opening became an even bigger extravaganza when Hammer arranged for John Denver to give a simultaneous concert in Kiev on behalf of the Chernobyl victims.

Our plan was to leave Moscow for Kiev in mid-morning, but at six A.M., the telephone in my hotel suite rang. It was Hammer. "Bob," he said, "I want you to stay by the phone. Something important is happening." So I sat and waited. Two hours passed. Then three. At nine o'-clock, Hammer called again and said simply, "Come to my office immediately."

At the Occidental offices, things seemed to be going crazy. Telephones were ringing; people were running around carrying boxes; something major was under way. The moment I arrived, a secretary brought me to Hammer's private sanctuary. Skipping the usual introductory pleasantries, he looked at me and said, "I can't tell you what's happening, but I want you to do something for me. I won't be able to go to Kiev today, and I want you to take my place. Go to Kiev; bring John Denver to the concert, and open the exhibit for me."

It didn't seem like a time for questions. If Hammer had wanted me to have more information, he'd have given it to me. So I said all right, went back to the hotel, packed, and flew to Kiev on a second Occidental plane with Tamar and John Denver.

Needless to say, the entire city was waiting for Hammer. I walked off the plane and announced to a huge delegation of public officials, drivers with sirens, and children with flowers that he wasn't coming, and they were gracious but obviously disappointed. Fortunately, I'd seen the exhibit enough times that I could speak intelligently about it. We drove to the museum, where I gave a brief speech emphasizing the bravery of the people of Kiev in the face of Chernobyl, how sad Hammer was that he couldn't attend the opening, and how sharing things like art was important to us all. Then the Ukrainian Minister of Health presided over a ceremony inducting Hammer and myself into the historical society of the Ukrainian

Medical Museum, and John Denver performed to an enthusiastic reception at the Kiev Concert Hall. It wasn't until I left the Soviet Union and picked up a newspaper at the airport in Zurich two days later that I learned what had happened. Immediately after I'd left Hammer's office, he'd gone to a hospital in Moscow to meet with David Goldfarb.

Goldfarb is a geneticist, a Jew who had protested Soviet violations of the Helsinki human rights accord. Afflicted with failing eyesight and diabetes, having lost one leg at the siege of Stalingrad during World War II and suffering from gangrene in the other, he'd tried to emigrate from the Soviet Union to the United States, but his exit visa had been denied. To this day, I don't know how Hammer did it, but he managed to put Goldfarb onboard the Occidental jet and fly him to New York.

Going to Kiev to cover for Hammer was one more step in my being regarded by the Soviets as someone more than a doctor who had treated the victims of Chernobyl and would soon disappear. It continued my evolution as a political and social figure in the Soviet Union, and that was important, because new opportunities were on the horizon. There was more work to be done.

CHAPTER

13

The Russian winter. Inevitably, I returned to Moscow in the snow. It was a different world from the one I'd seen before. Beautiful, white, painfully cold. The sun was bright; temperatures hovered at zero degrees Fahrenheit during the warmest hours.

As a Southern Californian, the only warm clothes I own are ski clothes. Victor took one look at my down jacket, realized I'd freeze to death, and arranged for more appropriate garb. Soon, I was wearing a fur hat, the kind anyone would look silly in anyplace but Moscow, but there it was natural. It wasn't a souvenir; I needed it to keep warm.

Saturday, January 25th, the day after my arrival, Baranov and I walked around the city. Red Square was wonderful, covered with snow. I was in Russia—forbidding and dangerously cold. More than before, I understood the role weather played in the timing of war, and why Napoleon and Hitler had failed when they invaded Russia.

Late in the day, I met Phil Donahue, who was in the Soviet Union taping a series of television shows. Chernobyl was on his itinerary, and the Soviets okayed his request that I accompany him to the power station. Because of the weather, air travel was impossible, and we met at Moscow Station for an overnight train ride to Kiev. The tracks had an eerie nineteenth-century glow. If Anna Karenina had appeared to hurl herself in front of an oncoming train, I wouldn't have been surprised.

Phil Donahue and I had our own compartments. The Soviet and American camera crews were bunched together in tighter quarters. Several of the cameramen had brought beer and I had a bottle of brandy, so once the train started rolling, we had a pretty good time. Donahue was interesting, asked intelligent questions, and had done his homework well.

Around midnight, we went to bed. My compartment was comfortable, but I couldn't sleep. There was too much drama, passing through Russia in the dead of winter. Outside, everything was black except for the falling snow. It was impossible to see more than several yards on either side of the train. My mind drifted back a mere forty hours to when I'd been swimming in Los Angeles in our backyard pool. Periodically, the train stopped to pick up passengers in small towns; God knows where.

At eight A.M., we arrived in Kiev, where Mikhail Shandala and Anatoly Romanyenko were at the station to greet us. They took us to a hotel for breakfast, then out onto the road to Chernobyl. Many of the trees in the evacuation zone had turned brown. Radiation coupled with the heavy spraying of decontaminants from the air had taken its toll.

Donahue had a very specific itinerary planned. He wanted to go to the power plant and start filming immediately. Obviously, he didn't understand the Soviet mind. First, they took us to a new settlement on the perimeter of the evacuation zone where workers who monitored the power station were now living. Then we stopped at a small Ukrainian village that looked like a set from *Fiddler on the Roof.* By now the day was wearing on, and Donahue was afraid we wouldn't reach Chernobyl until after sunset, when it would be too late to film. I tried to explain that no amount of screaming and jumping up and down would make the Soviets move faster, but he was apoplectic.

Looking at the positive side, I figured it was in the Soviets' best interests to give Donahue his Chernobyl show. Phil Donahue standing outside the power station, being seen by millions of people around the world, would be a signal to everyone that the crisis was over. So I advised him to stay calm. We'd get there, but not before the Soviets wanted us to.

We reached the power station in mid-afternoon, the first time I'd been on-site at ground level. The Mayor of Pripyat was there to greet us, and it occurred to me that he'd have a hard time winning reelection, since his constituents had all moved. Then I did something which may have been foolish: I went inside the power station. I was curious; I wanted to see what was there. The damaged reactor had been entombed, but Units 1 and 2 were producing power again, and I assumed it was safe for me to go.

A contingent of workers met me at the front gate. As a precautionary measure, I put on coveralls, boots, a special coat, hat, and mask—the same attire they wore. Inside, the power plant resembled a hospital without windows. The control room was a model of modern technology, with extensive monitoring of the Unit 4 reactor core.

The workers knew me as someone who had treated their colleagues in Moscow and were particularly friendly, talking at length about the difficult nature of their jobs. They were living away from their families in two-week shifts, the maximum number of consecutive days allowed in the evacuation zone. None of the television crew members came inside the plant with me, nor did Donahue.

That was significant only because a rather extravagant buffet lunch had been set out for us and, since I was the only guest present, the plant director and I ate lunch for twelve. Then I went back outside for a taping with Donahue, during which both of us stood on a mound of snow so the cameras could catch the power plant in the background. What the edited tape didn't show was that the mound was quite slippery, and both of us fell off into a ditch while the cameras were rolling.

The following morning, after an overnight train ride back to Moscow, I visited Tarmosian and Palamarchuk at Hospital Number 6. Both of them had recovered from their radiation exposure and were at the hospital for routine check-ups. Baranov and I spent much of the day reviewing miscellaneous medical matters. Then Victor took me to The Outdoor Pool in Moscow.

I went swimming—outdoors. I'd seen the pool in summer, and been told it was heated during the winter months. Being crazy, I'd gotten it in my head to give Russian-winter swimming a try if the opportunity arose. Victor and I walked to the pool. I'd brought my bathing suit and changed in the locker room, then ran outside and jumped into the water, which was warm. The air above was minus five degrees, and the pool was jammed. The differentiation in temperature between water and air was so great that I could only see two or three feet ahead of me through the rising steam. Overall, the experience was a cross between watching a Fellini movie and bumping into people.

The following day, January 28th, I returned to Los Angeles. Since then, I've traveled to the Soviet Union several more times. Meanwhile, like the rest of the world, I've watched as the Soviets put on trial and convicted a half-dozen power plant officials for safety violations in connection with the explosion. I've taken note of the fact that of six Chernobyl units originally planned, two will never be built and one more—the damaged reactor—will never function again. Still, for obvious reasons, I think of Chernobyl in personal terms.

Why did the Soviets invite me to begin with? The most ready explanation is they needed medical assistance and I was offering expertise, equipment, and supplies. But there are other possibilities, which aren't mutually exclusive. They might have envisioned that my being in Moscow would bolster their credibility at home and abroad. Remember, when I was invited, no one knew what was happening or how widespread the disaster was. By having me at Hospital Number 6 and putting me before the world press corps, the Soviets enlisted a credible outsider who could vouch for their version of what was going on.

Why do they keep inviting me back and promoting me as a "hero"? I guess that relates to how I handled the situation. I was flexible; I was willing to come back; I worked with rather than against them. And my presence gives them the opportunity to say, "Look, we don't have trouble with the American people. We get along fine with Hammer and Dr. Gale. It's the American government we quarrel with." Also, the Soviets might simply be trying to develop another private-sector American contact. Hammer is eighty-nine years old. He won't be around forever, and once he's gone, it will take a lot of people to fill his shoes. Whatever their motives, though, I'm sure the Soviets added up the pluses and minuses very carefully before making any decisions about me and will continue to do so.

There have also been gains for the United States as a consequence of my being in Moscow. I think it helps to have an American on Soviet television being perceived as a friend by the Soviet people. And beyond that, my being there gives us access to important medical and scientific data. There was a second private-sector offer made to the Soviets after Chernobyl. The Electric Power Research Institute in Palo Alto, California, has considerable expertise in nuclear safety matters and possibly the best non-governmental staff in the world. They offered aid immediately after the explosion. And even though the Soviets had put out feelers in West Germany, Sweden, and Switzerland for information on how to extinguish a radioactive fire, the institute's offer was declined. That might sound awful on the Soviets' part, but turn it around. Would the United States swallow its pride and share our own nuclear technology by accepting Soviet assistance in a comparable crisis? At Three Mile Island, there was fear that a hydrogen bubble inside the reactor was going to explode; that there would be a meltdown. If the Soviets had said, "We want to send scientists to study your American nuclear reactors; let us help you," our answer would have been no. Indeed, judged in that light, it's extraordinary the Soviets accepted me at all.

Why did I go? To save lives. I'm a physician; my role is to help people. I had expertise, so I offered it. I believe in medical care for injured people. Several years ago, the Israeli Air Force bombed a nuclear reactor in Iraq. I, like many Americans, thought it was necessary to preclude Iraq from developing nuclear weapons. But if there had been a release of radiation by virtue of the bombing and two hundred Iraqis were hospitalized, I'd have been willing to go there to treat them. Improving the level of health care in any country is good. Life is precious. Chernobyl's victims were deserving of the best medical care possible, regardless of their country's political system.

There's no way to know how many patients lived because of our medical intervention. Some would have survived without care in any hospital. Others surely would have died without extensive treatment. However, a few numbers might prove helpful. Approximately five hundred people were hospitalized for significant radiation exposure in the aftermath of Chernobyl. Of these, 203 developed radiation sickness of sufficient magnitude to require extensive medical intervention for bone marrow failure, burns, organ breakdowns, and comparable problems. We were powerless to do much for the most severely injured. Of twenty-three patients exposed to more than 600 rads of radiation, only one survived. And during autopsies, we learned that the damage caused by radiation was more complex than had been thought before. We didn't expect liver damage, but we found it. Gastrointestinal tract damage was spotty, which led us to believe it might have been caused by the ingestion of radioactive particles, which in turn radiated victims internally. On the other hand, we also learned that humans can survive much higher doses of radiation than previously thought possible.

Of the 203 patients treated for severe radiation sickness, 174 lived and 29 died. Two more victims had died immediately after the explosion, so the death toll was 31. Of the 174 who were treated and survived, perhaps half would have died without medical intervention, but that's a guess. The only way to get a precise statistic would have been to take all 203 patients, arbitrarily split them into two groups, treat half, and leave the other half on their own—an unspeakably cruel experiment that no one could condone.

All totaled, there were nineteen transplants. Six of these involved fetal liver cells, and each of those patients died of burns. The Soviets also performed six bone marrow transplants that had been planned before I arrived, and none of these patients survived. Seven more transplants were conducted by what I think of as the joint Soviet-American team. From this group, Tarmosian and Palamarchuk are alive today. Transplants were performed only on patients who were certain to die without them, and many of the transplant patients who did die expired for reasons other than bone marrow failure or graft-versus-host disease.

The thirty-one recorded deaths from Chernobyl are a sad loss, but small in proportion to what will follow. Best estimates, based on data provided by the Soviets themselves, indicate that over the next fifty years as many as fifty thousand people worldwide may die of cancer as a consequence of Chernobyl. Other possible consequences include birth defects and genetic abnormalities. Virtually all of these tragedies will be statistically undetectable. In the Western Hemisphere alone, six hundred million people will die of cancer during the same period, and there will be one hundred million cases of genetic disorders.

Overall, I'd say the Soviets responded well to the crisis. Clearly, their initial announcement regarding the accident was late in coming, and they were fortunate that prevailing winds carried the radioactive plume away from population centers rather than toward Kiev. However, beyond that, they handled the situation reasonably well, particularly after they realized the rest of the world could be trusted to deal responsibly with the situation. Much of the credit for the generally constructive Soviet response goes to Mikhail Gorbachev. He has a flexibility and dynamism hitherto unknown in Soviet leaders, and his policies, while not without problems, offer hope for the future.

Strange feelings come to mind when I look at Gorbachev and think about how much money our respective societies spend building weapons to destroy each other, particularly when I contemplate the presence of people like Baranov on the other side. It's sad, really. We as Americans don't know the Soviet people. We know the Soviet Union as a Communist country, but very few of us have considered who lives there apart from the context of their political system. For that reason, when people ask what was most important about my mission to Moscow, I'm inclined to answer: breaking down barriers. Barriers between the American and Soviet people, between an Israeli scientist and his Soviet counterparts, between the Communists and our world.

I'm not a Pollyanna. Throughout my stay in Moscow and Kiev, I knew I was in the Soviet Union. The mood was restrictive, and I never felt entirely free. I also assume that some of the people I interacted with had dual roles, and that their medical or scientific positions went hand in hand with activity on behalf of the KGB. The Soviets might not like my saying that, but I haven't entered a popularity contest in the Kremlin. From the start, I've been forthright and honest with everyone. So I'll continue saying what I think, and if at some point the Soviets decide my being there isn't in their best interests, I'm sure they'll let me know.

As for my long-term future, I can't solve the problems of nuclear energy or unilaterally end the Cold War. But I am absolutely determined to cure leukemia.

I'm a physician. I practice medicine. I do clinical research and take care of people. When all is said and done, I like wandering around a hospital late at night, studying X-rays, dealing directly with patients and helping them. I look at leukemia, and I say to myself, this illness shouldn't be. No one should die from this disease.

Everything in science has its proper moment. Fifty years ago, the knowledge and technology were becoming available to make an atomic bomb. Now they're available to cure leukemia. It's very tantalizing; we have the tools. If we couldn't cure anybody, we'd think it was hopeless— and that was once the case. But now we cure some people and we know more can be accomplished, so the rest becomes a technical matter.

We're going to cure leukemia. Setting high goals is dangerous because sometimes you fail, but it has to be done. What I'd like most is for someone to give me a large chunk of space and money, and tell me, "Do the job!" I'd set up my own unit, take ten very good people—I know who the best players are—and in five to ten years we'd have a cure. That would be the most satisfying thing I could do with my life.

As for Chernobyl and how it will affect my future, certainly the experience has forced me to think in broader terms. As a doctor, I've confronted my own mortality many times. But at Chernobyl, I confronted the spectre of universal suffering, mass death, and, most frightening, the mortality of all mankind. It's now scientifically possible in a worst-case scenario for nuclear weapons to eradicate all human life on Earth. For a long time, despite urging to the contrary, I've been silent on these issues. There isn't a nuclear power plant in the country that I haven't been invited to inspect by management or lie down in front of by antinuclear activists, nor a nuclear weapons rally—pro or con—that I haven't been asked to attend. Now I'm ready to speak out.

My thoughts, and those of my collaborator, Thomas Hauser, follow.

P · A · R · T
FOUR

The Final
Warning

CHAPTER
14

When primitive man first harnessed fire, a terrifying danger was involved. Out of control, fire could destroy the small forest around him—which, from his perspective, constituted the entire world. Over the millennia, our awareness of Earth's parameters has grown, but in relation to modern technology, the world we live in remains quite small. A radioactive cloud doesn't know when it crosses the border between the Soviet Union and Sweden. It just keeps going. Problems like acid rain and depletion of the ozone layer threaten us all. If nothing else, Chernobyl has forced us to contemplate these issues, to acknowledge that modern technology is a potent force and that when something goes wrong, it's an international event, not a national one.

This scientifically imposed unity among nations is particularly evident in nuclear matters. Once, if a hydroelectric dam burst on another continent, we could sit back and say, "That's their problem." But if a Soviet nuclear reactor explodes, it's our problem too. We have a substantial stake in how safely the Soviets manage their nuclear power program; they have a comparable interest in ours; and we both have cause to be concerned with reactors in other countries, particularly in politically unstable and underdeveloped areas of the world.

Everyone agrees that nuclear technology offers benefits, poses dangers, and represents an enormous challenge. However, balancing these considerations is a delicate matter. Once, wood was the world's primary fuel, but as populations increased and forests were cleared, coal became a necessity. Coal, in turn, was supplemented by oil. However, the reality of the situation is, our oil supply is finite, and coal causes significant environmental damage when burned. Thus, proponents of nuclear power argue that uranium is the basic fuel upon which future generations must rely and nuclear technology is the only answer.

Still, no other technology has remained under attack as unsafe and unsound with such persistence for so long. Automobiles cause massive pollution, and fifty-five thousand Americans die on the roads each year, yet no one calls for the abolition of cars. In December 1984, toxic gas leaking from an American-owned insecticide plant in Bhopal, India, killed two thousand people and seriously injured thousands more, yet no one suggested shutting down chemical plants around the world. Nuclear power is safe, say its supporters, and the future should be spent finding ways to make it safer, rather than urging that it be banned.

However, critics of nuclear energy adhere to a different view. They contend that nuclear power has been imposed upon the American people and ultimately the world by a small group of government officials, reactor manufacturers, and utility company executives with inadequate safeguards and a callous disregard for life. These critics distrust an industry which refers to accidents as "incidents" and disasters as "abnormal occurrences." They note that the growth of nuclear power was based in part on the optimistic assumption that scientists would find solutions to complex safety problems, but instead scientists have found more problems and few cures. To these critics, nuclear power is the most ill-advised industrial venture of all time, a "technological Vietnam" that continues to exist only because electric utilities are locked into the process by huge capital expenditures and the many years it would take to correct the error of their ways.

Regardless of the position one holds, it's clear that the issues surrounding nuclear power are not just a subject for theoretical research and debate. Chernobyl engendered more widespread political, social, and economic aftereffects than any industrial accident in history. Public support for nuclear energy has dropped in its wake. Many people have come to fear nuclear power more than is warranted by the facts, while others continue to fear it too little. In the United States, positions have hardened along partisan political lines, with liberals generally opposing nuclear energy and conservatives supporting it. However, in many parts of the world (including the Soviet Union), the political left advocates nuclear power, and the issue is best viewed apart from partisan political considerations.

What can be said politically is that in recent years public policy has become increasingly technological. Environmental control, "Star Wars" defense systems, and the like have complex scientific bases, and often this intimidates people. Too many of us feel we can't possibly fathom nuclear issues and, as a result, we don't think about them at all. However, in a democracy, people are obligated to become informed. Then we vote, and abide by the will of the majority. This is a responsibility all of us face, and it must be fulfilled, not on a partisan political or emotional basis, but on the basis of knowledge.

Neither of us—the authors of this book—has special expertise in nuclear matters. One of us is a medical doctor, the other is an attorney and author. However, we've done our best to educate ourselves on the issues involved.

We speak simply as two citizens in a democracy. We urge others to formulate their own views and make them known. And we repeat: No one need fear participating in this process simply because he or she doesn't fully understand the technology involved. For as Albert Einstein once observed, "Not even scientists completely understand atomic energy; each man's knowledge is incomplete."

Let's start with a simple fact: The world uses nuclear energy. Six months after the accident at Chernobyl, there were 374 nuclear reactors operating in thirty countries around the world. Another 157 were under construction, with 116 more being planned. These reactors account for 16 percent of all electricity generated globally, and the demand for electricity is increasing radically worldwide. Thus, no one country's decision regarding the merits of nuclear power will alter the reality that, whether we like it or not, we live in a nuclear age.

In the United States, as of this writing, there are 102 nuclear power plant reactors licensed to operate and 24 more under construction These reactors provide 17 percent of our nation's annual electrical output—a figure expected to rise to 20 percent by the early 1990s. Six states derive more than 50 percent of their electric power from nuclear energy, led by Vermont with 71 percent. In terms of capacity and kilowatts generated, our nuclear power program is far and away the largest in the world.

However, the idea that nuclear power will spread through the United States as inexorably as electricity blanketed rural areas a half-century ago is long gone. No nuclear power plant reactor has been ordered from a manufacturer since 1978, and every unit ordered since 1974 has been canceled or indefinitely postponed. Our nuclear generating capacity will peak in 1992, and then decline as the first generation of nuclear plants grows obsolete and is taken out of service. For two decades, the United States has been the major supplier of fuel, equipment, and technological expertise in the growth of nuclear energy worldwide. Now, many people believe that, having led the world into the age of nuclear power, the United States may well lead it out.

However, the United States is no longer the sole parent of nuclear technology. As of today, many countries have competent nuclear physicists and advanced nuclear power programs of their own. France derives 70 percent of its electricity from nuclear power; Belgium, 60 percent; Sweden, 50. There are thirty-four nuclear power plants operating in Japan and thirty-eight in the United Kingdom. The Soviet Union, despite Chernobyl, has announced plans to quintuple nuclear power production by the year 2000.

Also, the Third World is discovering nuclear energy. The average Third World country consumes 400 kilowatt hours of electricity per person annually, compared to 9,600 in the United States. Approximately 1.7 billion people in developing areas have no electricity at all. Many of these countries find themselves without alternatives to nuclear power. They have no reserves of oil or coal, and little foreign currency to pay for fossil fuel. They're going to use nuclear energy whether we do or not.

What are the dangers, at home and abroad? First and foremost, there's the risk of an accident. Our society has enormous faith in technology, but nuclear power systems stretch that faith to its limits. They're extraordinarily complex, with thousands of pumps, valves, pipes, circuits, and motors. Problems can arise from errors in design, manufacture, installation, maintenance, operation, and external sources like floods, fire, earthquakes, and tornados. Corrosion, vibration, stress, and simple deterioration due to age can result in a single small defect. One defect can lead to another. The second can cause a third, which no computer can foresee or control. Most machines are allowed to run until they break and then we fix them, but nuclear reactors can't operate on that basis.

It may be that a "Chernobyl-type accident" can't happen in the United States. The Chernobyl unit had a Soviet design, which used graphite as a neutron moderator and boiling water as a coolant. It also lacked a complete containment structure, and incorporated a positive power coefficient feature which many analysts believe contributed to its instability. However, as observed earlier, when it comes to nuclear matters, Soviet accidents are our accidents; and other types of reactor accidents can happen here. Many safety features of American units have been "tested" only by mathematical simulation, not empirical data. In some instances, utilities have been reluctant to acknowledge and correct defects in plants under construction because to do so could be construed as an admission of safety flaws in existing systems. And then there's the most fallible component of any nuclear power plant—the men and women who operate it.

People get tired; they have bad days; they act without thinking and make mistakes. To be certified as a control room operator, an applicant must pass a test administered by the Nuclear Regulatory Commission with a score of 80 percent. One wonders if 80 percent performance in a nuclear reactor control room is truly satisfactory, and how these operators learn the other 20 percent of information essential to their job. At some utilities, operators are chosen on the basis of union contracts which place seniority ahead of competence. Too often, they just don't do their work properly. At nuclear power stations across the country, federal investigators have found control room operators asleep on the job. At the Portland General Electric Trojan Station, a control room operator was listening to a basketball game piped into the telephone at his control console while radioactive water was overflowing from a tank and flooding the adjacent building. At Brown's Ferry, an inspector entered the control room and saw no one on duty, even though the reactor was at full power. One operator had gone to the men's room, the other was behind the instrument panel. The use of alcohol and on-duty drug abuse by control room personnel are reflected in numerous federal reports.

Major accidents have humble origins, and once they begin, systems break down. At Three Mile Island, the control room had elaborate panels with hundreds of lights designed in theory to provide operators with information necessary to diagnose problems and carry out a proper emergency response. However, in practice, the panels were so spread out and their printed messages so small that the operators could read only a small portion from any given position at one time. Also, the lights weren't grouped to distinguish urgent problems from lesser ones. Thus, when the crisis began, more than a hundred red, green, white, amber, and blue warning lights glowed like Christmas decorations around the room while a single alarm sounded and the operators didn't know where to begin. "I would have liked to have thrown away the alarm panel," one of the control room operators said later. "It wasn't giving us any useful information."

At Three Mile Island, the scientists lost control. And even though they recaptured it, close to a billion dollars and the efforts of several thousand technicians were required to decontaminate the area and shut the reactor down. More than a year after the accident, the damaged unit was relying on safety systems that had been operating without maintenance for periods longer than had ever been planned. Fifteen hundred gallons of water were being added daily to replace leakage into the containment sump. With control rods having melted in the intense heat, the plant's overseers were dependent upon boron-laden water being pumped through the reactor cooling system to keep the enriched uranium fuel from going critical again.

In sum, accidents happen. This is why the nuclear industry continues to insist upon laws limiting its liability for damages arising out of nuclear accidents. And beyond chance occurrences, there is the possibility of terrorism and sabotage aimed at a nuclear reactor.

The prime example of nuclear sabotage in the United States occurred in 1961 at an Atomic Energy Commission reactor-testing station in Arco, Idaho. Three technicians were on a four P.M. to midnight shift. The reactor had been shut down, and the technicians were to lift the central control rod by four inches—enough to perform certain maintenance but not so far that the reactor would start. However, one of the technicians lifted the rod out completely. A chain reaction began, power surged, and the reactor exploded. All three workers were killed, one of them impaled on a control rod that ran through his groin, pinning his body to the ceiling. Ten years later, an Atomic Energy Commission memorandum revealed, "The accident . . . is now known to have been initiated on purpose by one of the operators bent on murder-suicide." The perpetrator apparently was distraught because he thought his wife was having an affair with one of the other operators on the same shift.

Clearly, control room operators have the knowledge, means, and opportunity to effectuate a crisis. What sort of psychological monitoring, if any, exists? What is to prevent a control room operator from going mad? Nuclear Regulatory Commission regulations mandate that nuclear power plants provide protection against sabotage by one insider but no more. Is that protection enough? And what about attacks from the outside? A nuclear power plant would make an extraordinary target because of its potential for the release of radioactive material. Yet federal regulations only require protection against sabotage by up to three external attackers operating as a single team with weapons that can be carried by hand. Reactors need have only five security guards on duty at any time.

Proponents of nuclear power argue that sabotage and terrorism are of minor concern. After all, many groups have the technical expertise to kill thousands of people and don't. "If killing fifty thousand people is their desire," writes one observer, "there are many easier alternatives for accomplishing it. They could release poison gas into the ventilation system of a large building; dynamite the structural supports in a sports stadium so as to drop the upper tier down on the lower tier; discharge a large load of napalm or perhaps even gasoline on the spectators in a sports stadium; blast open a large dam; poison a city water supply."

The above constitutes an imaginative list, but nothing on it would render an area permanently uninhabitable. Nor would any of the suggested terrorist tactics require the immediate evacuation associated with a potential nuclear disaster.

Evacuation requires moving people from high-risk to low-risk areas, using distance as a defense. The term implies an orderly planned exodus, but in the aftermath of a serious nuclear accident in the United States, and particularly near a large city, more likely there would be panic and chaos.

How do you evacuate? Who will run the buses and trains? Should pregnant women and children be moved first? What about hospital patients? Should medical personnel be asked to stay behind to care for the bedridden if the general population is moved? How do you empty a prison and control prisoners in the aftermath of a nuclear power plant accident?

Because of our fortunately limited experience with nuclear accidents, our contingency plans are based substantially on theory and guesswork. However, these plans assume that emergency workers who live within the danger zone will report for work whether or not their own families have been evacuated, and that workers safely outside the zone will enter the area to report for work. Military personnel are likely to do so. Police officers, doctors, and paramedics might. But a large number of people—bus drivers, tow-truck operators, communications experts, check-point monitors, traffic controllers, and others—are likely to ignore their obligations. In the Soviet Union, by the Soviets' own count, 177 Communist Party members failed to carry out their duties during the evacuation of Pripyat, and rumors abound of others complying only out of fear of being shot. Given the less authoritarian nature of our society, what sort of compliance could we expect? On any normal day, the roads in and around most American cities are jammed for hours. Would they really be passable in a nuclear crisis? Does anyone seriously think we could evacuate Los Angeles or New York?

These are some of the questions one must ask regarding aberrational incidents like sabotage and nuclear accidents. Moreover, the normal everyday functioning of nuclear power plants involves certain risks, foremost among them the dangers inherent in the nuclear fuel cycle. This cycle begins with mining as uranium ore is extracted from the earth. Then comes milling—a process by which ore is crushed, ground, and treated chemically to extract uranium in a form called "yellowcake." Next, the yellowcake is converted to a gaseous compound, and enriched to increase its percentage of uranium-235 isotopes. Finally, the compound is converted to uranium oxide pellets and packed into fuel rods. Each of these steps involves the release of above-normal levels of radiation. However, far and away the most troubling aspect of the nuclear fuel cycle is the matter of nuclear waste. By 1985, twelve thousand tons of highly radioactive spent fuel had been produced by nuclear power plants in the United States. According to the National Academy of Sciences, by the year 2000, seventy-two thousand tons will have been brought into existence. Moreover, in addition to these high-level wastes, a far greater volume of low-level wastes such as mill tailings and by-products of the enrichment process has accumulated. Critics of nuclear power point to an Environmental Protection Agency estimate that by the year 2000, there will be one billion cubic feet of low-level nuclear waste in this country—enough to cover a four-lane highway coast-to-coast one foot deep. Proponents counter that all the spent fuel produced to date by commercial nuclear power reactors in the United States could be stacked on a single football field three feet deep. However, more important than football fields and highways is the fact that we don't know what to do with nuclear waste.

High-level radioactive waste cannot be destroyed, only guarded, and must be isolated from the environment for tens of thousands of years before it's safe. At present, we simply don't know how to do this. No human society has remained in existence for tens of thousands of years, so it's necessary to develop a disposal system independent of human control. No technical process has been able to operate without error for what is essentially forever, yet that's what's involved in sequestering nuclear by-products.

Over the years, a variety of suggestions have been advanced: launching rockets loaded with nuclear waste into space (but what if a rocket explodes on launching?); burying containers in the polar ice caps, where their heat would allow them to burrow downward through the ice until they hit bedrock (but would this cause the polar ice caps to melt?); packing the waste in needle-shaped packages to be dropped into the ocean, where they'll penetrate thick clay sediment at four-mile depths (followers of Jacques Cousteau don't particularly like this idea); drilling shafts six miles into the earth to serve as burial chambers (but what if there's an earthquake?); storing the waste in salt or granite geological formations (but how can we ensure the structural integrity of the storage canisters?).

At present, spent fuel can be transported and stored in specially designed casks constructed of steel, lead, and various corrosion-resistant alloys. These casks have been laboratory-tested by putting them in trucks and crashing them into a seven-hundred-ton concrete wall at eighty miles an hour; dropping them from a height of two thousand feet; putting them on railroad tracks and running a locomotive into them at ninety miles per hour; and submerging a cask in a pool of burning aviation fuel. In each instance, the container survived, but none of these tests involved the passage of time, and the United States Department of Energy estimates that no canister filled with highly radioactive nuclear waste can be expected to resist corrosion for more than several hundred years.

Thus, at present, most spent nuclear fuel is "temporarily" stored in steel-lined pools adjacent to nuclear power stations, and the potential for damage continues to mount. Indeed, within the scientific community, it's assumed that at least one major nuclear waste accident has already occurred. During late 1957 or early 1958, a huge explosion destroyed what is believed to have been a Soviet nuclear waste disposal site at Kyshtym in the Ural Mountains. According to Zhores Medvedev, a biologist who later emigrated from the Soviet Union to London, "There was an enormous explosion, like a violent volcano. The nuclear reactions had led to an over-heating in the underground burial grounds. The explosion poured radioactive dust and materials high up into the sky. Tens of thousands of people were affected, hundreds dying. A large area, where the accident happened, is still considered dangerous and closed to the public."

Perhaps Kyshtym is where Angelina Guskova and her Soviet brethren gathered their knowledge of beta burns and radiation scoring. Perhaps not. But the fact remains that with almost four hundred commercial nuclear power reactors in operation around the world, no long-term nuclear waste disposal system exists. The people charged with managing nuclear programs profess confidence that they'll be able to develop a perpetually closed system. However, they've yet to do it.

Moreover, in assessing the everyday dangers of nuclear power, it should be remembered that "atoms for peace" programs spread the technological expertise and raw materials necessary to fashion nuclear weapons. Weapons grade plutonium can be separated from spent reactor fuel at a cost well within the budget of many Third World countries. When India followed the United States, the Soviet Union, Great Britain, France, and China into "the nuclear club," it relied on materials and technology supplied by the United States and Canada as part of its "peaceful" nuclear power program. More recently, Pakistan appears to have done the same.

In theory, the International Atomic Energy Agency is charged with preventing the use of nuclear power technology and fuel for arms production. However, like most multinational agencies, the IAEA is unable to effectively sanction nations that violate its safeguards. Indeed, after the 1981 Israeli bombing of Iraq's nuclear reactor, an IAEA inspector testified before Congress that host nations have the right to veto agency inspectors on the basis of nationality, limit the inspection of reactors under construction on their soil to once every four months, require advance notification of each inspection, and declare facilities off-limits to inspection by simply stating that they contain no thorium, uranium, or plutonium—materials essential to converting nuclear plowshares into swords. In sum, the safeguards aren't safe.

These, then, are some of the dangers inherent in nuclear power. However, it solves nothing simply to take a stand against a particular energy source; in this case, uranium. The United States uses one-third more electricity today than just before the 1973 Arab oil embargo, and every indicator suggests that this trend will continue. Thus, opponents of nuclear power must stand for something and offer alternatives.

What are the choices? First, there's hydroelectric power—using dams and falling water to generate electricity. Hydroelectric power played a significant role in America's industrial development, and filled 30 percent of our electricity needs as late as World War II. It's cheap, renewable, and largely untapped in Asia, Africa, and Latin America, where less than 10 percent of the potential is exploited at present. But in the United States, most hydroelectric sites have been fully developed, and additional dams would threaten serious ecological damage.

Geothermal power comes from natural heat, trapped underground in the form of hot rock, hot water, and steam. Like hydroelectricity, it constitutes a renewable resource, but there the similarity ends. Geothermal systems operate at extremely low thermal efficiency, require extensive space, and at best can be expected to satisfy only a tiny fraction of any nation's electrical needs. The same holds true for oceanic thermal power, which relies on the temperature difference between surface and depth waters in tropical areas.

Tidal energy, wind, and wood offer little hope for the future. Natural gas as an energy source is constricted by availability and cost. Bioconversion—the transformation of municipal waste to energy—holds some promise, particularly when one considers it as auxiliary to the disposal of unwanted garbage. However, the heat content of municipal waste is low, the cost of conversion is high, and even bioconversion's most ardent supporters agree that it will never provide more than 2 or 3 percent of any nation's electrical power.

Next comes coal—the source for approximately 56 percent of all electricity generated in the United States. Coal is cheap and plentiful. We have, within our borders, 280 billion tons of recoverable reserves, enough to provide energy for hundreds of years. However, burning coal poses serious environmental hazards in the form of acid rain and the release of carbon dioxide into the atmosphere, leading to a global "greenhouse effect." Coal miners, on average, live three years less than their counterparts in other industries, and there's no way to know how many deaths in the population-at-large result annually from pollution caused by coal-fired turbine generators.

Oil is "cleaner" than coal but more expensive and less plentiful. Prior to World War II, the United States produced enough oil to meet all of its domestic needs, with some left over for export. However, since then, domestic consumption has risen dramatically, OPEC has come into existence, and the OPEC nations have consolidated their power. Today, there are approximately 670 billion barrels of recoverable oil reserves in the world, and OPEC controls two-thirds of them. Indeed, in a single two-month period in 1973, OPEC raised the price of oil from $3.10 to $11.65 per barrel, and by 1981 the price was $35.00. Additional reliance on oil would mean increased dependence on foreign powers and possibly an increased risk of world conflict because of Middle Eastern political implications. And the Petroleum Age may be coming to an end anyway, inasmuch as it's unlikely that new oil discoveries will extend world resources significantly beyond what we already know exists.

In sum, every energy source has its limitations and price in terms of dollars, lives, and environmental damage. No source can be pursued to the exclusion of all others, and none can be eliminated without a satisfactory substitute being put in place. There's a trade-off between benefits and risk, and decisions regarding nuclear power must be made in this context. Still, each of us should imagine what it would be like to have a stranger appear at our door tomorrow and tell us we had to leave home immediately, never to return. That's what happened to 135,000 people at Chernobyl. Each of us must understand that the physical effects of Chernobyl will be with us, our children, and our children's children for generations to come. And above all, each of us must consider the words of Adlai Stevenson, who, in his last public address, reminded the world, "We travel together, passengers on a little spaceship, dependent on its vulnerable reserves of air and soil; all committed for our safety to its security and peace; preserved from annihilation only by the care, the work, and I will say the love we give our fragile craft."

Stevenson's thoughts go to the heart of the matter. For as horrifying as the accident at Chernobyl was, it in no way represents the ultimate nuclear power disaster. Chernobyl was fifty miles from the closest major population center; wind currents at the time of the accident were favorable; only 3 to 5 percent of the reactor's core was released; as callous as it may sound, the Soviet Union is a country which can afford to evacuate a thousand square miles; and most important, Chernobyl was an isolated accident. By contrast, nuclear power as a whole threatens the ecological balance of the entire planet. Indeed, according to the Lawrence Livermore National Laboratory in California, the Chernobyl disaster will ultimately release one-third as much long-term radiation into the atmosphere as all the world's nuclear weapons detonations to date.

In a worst-case scenario, there may come a time when the effect of these nuclear power by-products—from accidents and the everyday nuclear fuel cycle—inflicts irreversible damage upon our planet. And should this happen, the things we rely on for life will begin to die. At present, this is only theory. But often in the 1980s we've been reminded that our relationship with the Earth is changing. In 1985, scientists discovered a decline in the level of atmospheric ozone over Antarctica. The ozone layer is a precondition to life on Earth, protecting plants and animals from the sun's ultraviolet radiation. One year later, a 1986 study by British meteorologists demonstrated that rising atmospheric levels of carbon dioxide and other "greenhouse gases" had already led to a worldwide warming trend.

Nature, of course, plays cruel tricks of its own without human prompting. In August 1986, eighteen hundred people died when a cloud of lethal gas sprung from the bottom of a lake in Cameroon. Earthquakes, tornados, and other natural disasters are common. But as a general rule, rages of nature's making are accommodated by the environment. Flood waters recede; winds abate; life begins anew in the shadow of spent volcanos. By contrast, man-made disasters threaten permanent destruction. And it may be that by the time the full adverse effects of nuclear power on the environment are known, it will be too late to change direction.

All living things derive sustenance from the environment. Obviously, it's in our self-interest to protect it. Also, we as a generation have a moral obligation to satisfy our needs without diminishing the quality of life and prospects of survival for future generations. Thus, given what's known about the risks inherent in nuclear power, we must ask: Should dangers of this magnitude be allowed to continue?

It may be that, in nuclear matters, we've reached the limits of our technological expertise; that the problems of nuclear radiation will never be solved. That's hard to accept. All of us have been led to believe that scientists can resolve scientific dilemmas, but certain physical realities erect insurmountable barriers that can't be crossed. For example, we've come to accept the notion that dead people can't be brought back to life. That's a cold hard chemical matter. We can't "solve" death; we can only delay it. The dangers inherent in splitting atoms are rooted in the stark unyielding laws of physics. Perhaps they too cannot be solved.

These issues aren't pleasant to contemplate. No one likes to focus on death and disaster, but we need answers. For the Soviets, the present course is straight ahead. One month after Chernobyl, Victor Sydorenko of the State Nuclear Safety Committee declared, "The Chernobyl accident cannot and should not change the strategy of the nuclear industry. Proper lessons will be drawn, however bitter they might prove, and we will continue to advance. Progress cannot be stopped." Sydorenko's remarks were echoed by Valery Legasov of the Soviet Academy of Sciences, who wrote for *Pravda*: "I am profoundly convinced that nuclear energy stations are the pinnacle of achievement in power generation. Nuclear energy is the beginning of a new phase in the development of civilization. The future is inconceivable without the peaceful utilization of nuclear power."

In the United States, the response to a domestic accident comparable in magnitude to Chernobyl would probably be similar. After all, each day an estimated nine hundred Americans die from illnesses related to cigarette smoking. The typical one-pack-a-day smoker lowers life expectancy by 6.4 years for men and 2.3 years for women, yet we still smoke. Being twenty pounds overweight is more dangerous to the average person than current levels of nuclear fallout, but most Americans exercise too little and eat too much. Forty percent of all fatal accidents in this country occur at home. We drive in cars and fly in planes with an almost fatalistic awareness that death could lie moments ahead. So if a nuclear reactor in Middle America exploded tomorrow and thousands of people were killed, what would happen? Would we acknowledge that billions upon billions of dollars had been misspent and dismantle our nuclear power capacity? It's unlikely. But there are steps which can be taken now to deal with the present nuclear energy crisis, and we recommend the following:

1. *Better communication within the industry regarding design flaws, "minor" malfunctions, and safety lapses.* Eighteen months prior to the accident at Three Mile Island, a similar chain of events occurred at a nuclear power station in Ohio, but control room operators noticed the open safety valve in time to avert a crisis. Public utilities are reluctant to air their problems for fear of negative publicity. However, if Three Mile Island personnel had been aware of the Ohio incident, their own crisis might have been avoided. Recently, the nuclear power industry has taken positive steps in this area. More are necessary.

2. *Improved personnel selection and training.* Nuclear power is so fraught with risk that it needs the best people possible in every job. The immediate responsibility for managing a nuclear power accident belongs to the control room operators on duty when the malfunction occurs, whether it's at noon or four in the morning. Yet these operators are often inadequately trained to deal with emergency situations, and sometimes lack the basic skills essential to top job performance. "We have tended to train people to read cookbook recipes instead of becoming good cooks who understand everything in the kitchen," acknowledges William Lee, chairman and chief executive officer of North Carolina's Duke Power Company.

3. *No nuclear power plant should be built in the immediate vicinity of a large population center.* One could argue that the "danger zone" around a nuclear power station is the entire planet. But the area most vulnerable to fallout is the area nearest the plant, and evacuating a large city is virtually impossible. Scientific calculation and the public perception of risk have to be balanced. Consolidated Edison once applied for permission to build a nuclear plant in the Ravenswood section of Queens, adjacent to midtown Manhattan. Five and a half million people lived and worked within five miles of the proposed site.

4. *We need a national medical program for responding to nuclear accidents.* Obviously, prevention is better than treatment, but if treatment becomes necessary, the United States doesn't have a true national policy. In terms of dealing with small accidents where only a handful of people are injured, our preparation is good. Most nuclear power stations have an ongoing relationship with local hospitals, where people are trained in the care of nuclear accident victims. So in theory, if five people are brought to an emergency room, the hospital will know how to deal with them. But what if we have an intermediate-size accident with several hundred casualties? The resources to respond exist, but they're not properly linked and, with such an accident, normally simple things become quite complicated. Blood samples from the victims might be radioactive. Will we send them to the main clinical laboratory in each hospital where the patients are located, in which case the entire lab could become radioactive? Should radioactive urine be introduced into the pipe system of each hospital? How will we control the flow of friends and relatives who come to visit? And these concerns deal with mid-range casualties. If there's a reactor meltdown with thousands of victims, we simply wouldn't be able to deal medically with the situation.

5. *No more nuclear power stations should be built until a solution is found to the problem of nuclear waste.* We simply cannot afford to bring more and more of these lethal elements into existence without being able to dispose of them safely.

6. *Greater international cooperation and data sharing are necessary.* No one nation can preserve the ozone layer, end acid rain, stabilize global temperatures, or eliminate the risk of nuclear radiation. An international effort is required, and this will entail sharing data on all aspects of nuclear power to an extent greater than currently practiced through the International Atomic Energy Agency. When one of our reactors suffers a minor malfunction, do we tell the Soviets about it? No; nor do they tell us. But we should, and they should, because a nuclear accident anywhere is a nuclear accident everywhere. The sovereignty of nations cannot be allowed to compromise world safety. We have to be concerned with the integrity of nuclear reactors in other countries as well as our own.

7. *We need greater safeguards against the conversion of "peaceful" technology to weapons use.* Although nuclear power is sold to Third World countries under the slogan "atoms for peace," nuclear reactors are integral to the proliferation of nuclear weapons. Unless we can halt the use of this technology for weapons manufacture, we shouldn't export it.

8. *The development of solar power.* It's one thing to have an idea and quite another to implement it. One can't simply say "solar energy" and make it happen, but the sun is the most abundant source of energy available to us. It's infinitely renewable and non-polluting. At present, solar technology is limited by the need for an energy storage system (for times when the sun doesn't shine) and the fact that sunlight is diffuse—that is, it's spread out over the Earth's surface, making it difficult to gather the sun's power. But over the past thirty years, considerable progress has been made in converting solar energy to electricity through photovoltaic cells. We make the assumption that a society capable of landing a man on the moon and splitting the atom can harness solar energy. At the very least, a portion of the billions of dollars spent annually on nuclear power should be redirected to the quest for electricity from solar power and other innovative technologies that may evolve over the decades.

Nuclear energy—like space exploration, genetic engineering, and other technologies—is neither inherently good nor inherently evil. Indeed, further study may demonstrate that its long-term environmental consequences are less damaging in many respects than the use of coal, oil, and other fossil fuels. But nuclear energy must be made safer. Whether it can be made safe enough, though, is open to question; and thus, enlightened self-interest demands that we reevaluate our policies before it's too late. This should include attention to the development of inherently safe reactors reliant upon laws of physics rather than human intervention to eliminate risk.

As for Chernobyl, it may be that the greatest contributions made at Hospital Number 6 were not the lives saved but the lives lost. For the failure to save lives demonstrated how deadly nuclear power can be and how helpless the world is when radiation rages wild.

In the end, we all live near Chernobyl.

CHAPTER
15

In a 1931 address to students and faculty members at the California Institute of Technology, Albert Einstein warned, "Concern for man himself and his fate must always form the chief interest of all technical endeavors in order that the creations of our mind shall be a blessing and not a curse to mankind."

Einstein's words, spoken long before the making of the atomic bomb, remain true today. And while this book is primarily about nuclear power, we would be delinquent in fulfilling our responsibility as authors if we failed to discuss a larger issue. For nuclear power grew out of nuclear weapons; in many respects, they're similar. And Chernobyl is a chilling reminder of the ultimate risk to humanity today.

In past centuries, the primary causes of untimely death were famine and plague. During the mid-fourteenth century, the Black Death killed one-third of the population between India and Iceland. Ten million people perished in the Great Bengal Famine of 1769–1770. But in the twentieth century, man-made death has surpassed natural horrors. Fifty million people died in World War II. Since then, we've endured over a hundred international conflicts and civil wars, claiming ten to twenty million lives. And as awesome as these numbers are, a more frightening spectre looms.

Nuclear weapons are exponentially more destructive than their predecessors. We have missiles capable of delivering them to any spot on the globe. Given our ability and that of our adversaries to strike instantaneously, we face the grim reality that the Soviet Union could obliterate the United States in a matter of hours, and we could do the same to them. Thus, Richard Rhodes observes, "Though they bristle with holocaustal weapons, the superpowers confront each other totally vulnerable, totally dependent for their continued survival on mutual and reasonable restraint. The bomb, the final word on the accumulation of power—that matter properly arranged is all power—has saturated national sovereignty and shorted it out." David Inglis, a physicist at Los Alamos National Laboratory from 1943 to 1946 and past chairman of the Federation of American Scientists, concurs: "National security has ceased to exist in any absolute sense. Our awakening tomorrow to participate in the life of a happy and healthy nation is subject to the whim of a foreign totalitarian regime—a regime armed with H-bombs and presumably with the means to deliver them almost unhindered. That whim is fortunately kept in check by our ability to retaliate, giving us a perhaps illusory feeling of confidence that we will work and enjoy many tomorrows. But national security has been reduced to a question of probability, instead of certainty, that we will survive any year without being blown to bits."

Whether nuclear weapons threaten or, to the contrary, stabilize world peace has been a source of controversy for decades. Many argue that policies of mutually assured destruction actually engender global harmony, that a planet free of nuclear weapons would become dangerously safe for conventional war between the world's superpowers.

However, the price of nuclear peace has been the omnipresent threat of nuclear war. To date, we've avoided a holocaust, but the world's nuclear arsenals continue to grow. We defend ourselves with a strategy of deterrence based on mutual suicide, and remain dependent for peace upon weapons that all sides consider too terrible to use.

Meanwhile, the danger spreads as nuclear weapons proliferate like the metastasis of a deadly cancer around the globe. Britain joined the United States and Soviet Union in the "nuclear club" in 1955; France in 1960; China in 1964; India in 1974. South Africa has been able to produce nuclear weapons since 1981, and Israel has become an undeclared nuclear power, with an estimated one to two hundred nuclear warheads in reserve. Some nations—notably Canada, Japan, and West Germany—have chosen not to develop nuclear weapons, although they have the technology to do so. Others, such as Libya and Iraq, have aggressively sought to become nuclear powers.

Unfortunately, over the years, the barriers to nuclear weapons production have fallen. Many countries are increasingly competent in nuclear matters and, at present, the main technical obstacle to weapons production is obtaining an adequate supply of fissionable material. Einstein put the matter in perspective best in warning, "What nature tells one group of men, she will tell in time to any group interested and patient enough in asking the questions."

Clearly, the proliferation of nuclear weapons increases the risk of nuclear war. And beyond that, for as long as even one country possesses nuclear warheads, an accidental detonation could occur. Many people believe we can "live with the bomb." Since the 1960s, all of our nuclear weapons except those on naval vessels have been fitted with electronic locks requiring appropriate political authorization before firing. Because of potential communications problems, naval vessels can, under certain circumstances, fire on their own, but such action requires the approval of approximately thirty specified shipboard personnel. Indeed, some observers believe the United States has so many mechanical safeguards, procedural checks, and other restraints built into its nuclear arsenal that we might be unable to launch a nuclear attack even if our political and military hierarchy appropriately decided to do so.

One hopes that the Soviet Union and other nuclear powers are as determined to avoid the accidental use of nuclear weapons as we are. However, good intentions by no means constitute a fail-safe system. Accidents happen. Perfection is impossible, both in the machines we make and the people who run them. Who are the men and women with their fingers on nuclear buttons, for us and for the other side? Are they brilliant or of ordinary intellect? Happy or depressed? Do they drink too much? Smoke marijuana? Are they angry at their children, or upset because their spouse is having an extramarital affair?

And what of the technological safeguards? Bhopal, Chernobyl, and the Challenger space shuttle disaster all demonstrate the limits of technology. Against all logic and technological assurances, power blackouts occur. Civilian space satellites have already been labeled "the computer hacker's next frontier." Can we really be sure that military installations will remain inviolable? Is there a "defective booster seal" lurking at one of our missile silos, or at a silo in the Soviet Union?

Miracles seldom, if ever, happen. Catastrophes are far more common. At present, the United States armed forces have an estimated 722 military units with nuclear capability. Presumably, the Soviets have a comparable number. How safe are our respective nuclear triggers? Will everything continue to work the way it's supposed to?

The world can't stockpile more and more nuclear weapons without increasing the chance of a nuclear accident. And as stockpiles grow, the risk of intentional detonation also becomes more real. Each year increases the odds that a nation with irresponsible leaders or an unstable government will acquire nuclear arms. Motives ranging from self-preservation to religious fanaticism could inspire their use. Israel, Iraq, South Africa, Iran, and a dozen other countries might find "appropriate" circumstances to justify nuclear action. One nation, of course, has already found circumstances that warranted the use of nuclear weapons against an enemy's civilian population, and we think of that nation, the United States, as one of the most enlightened and humane on the face of the globe.

One hopes we're still living in the post-war age and not a new pre-war era. But in many ways, society is becoming increasingly violent and hatred is rampant around the world. There are no guarantees against nuclear weapons use, and even an anonymous nuclear attack is possible. Few of us think a nuclear warhead will be unleashed tomorrow. But none of us think we'll be killed in an automobile accident either, and each year on our nation's highways, fifty-five thousand of us are proven wrong.

What would be the consequences of a single nuclear explosion given the magnitude of today's bombs? The truth is sufficiently terrifying that most people banish it from their thoughts. However, it's important to understand what one nuclear warhead is capable of doing.

The bomb that destroyed Hiroshima had an energy yield of about thirteen kilotons (thirteen thousand tons of TNT). By contrast, most strategic nuclear weapons today are in the one-to-ten-megaton range (one to ten million tons). The largest nuclear device ever detonated was a fifty-eight-megaton warhead exploded by the Soviet Union in an atmospheric test in 1961. The largest United States test was a 14.8-megaton detonation in 1954. Putting these numbers in context, the energy yield of all the conventional bombs dropped by the Allied forces during World War II was approximately three megatons. The energy yield of all munitions expended by the United States in Vietnam was less than four megatons.

The energy from a nuclear explosion is released in many

ways. At the instant of explosion, "direct radiation" consisting primarily of gamma rays is emitted and, in the case of a one-megaton bomb, could be expected to kill unprotected persons within an area of six square miles. Almost simultaneously, electromagnetic waves are emitted in a single pulse which, while not harmful to humans, will take its toll. Studies indicate that the electromagnetic pulse from a single strategically placed one-megaton detonation would damage solid-state electrical circuits throughout the United States, wreaking havoc upon utility power grids, radio transmitters, and computer systems. Over the next few seconds, still more energy is released in the form of "thermal radiation"—heat—moving outward at a velocity slightly less than the speed of light. The thermal pulse of a one-megaton bomb would cause mass fires and potentially fatal third-degree burns to unprotected persons at distances of up to seven miles. Next comes the "shock wave," which, in the case of a one-megaton atomic bomb, is capable of destroying buildings five miles away and causing winds strong enough to turn otherwise harmless objects into lethal bludgeons and spears. And added to all of this is the peril of radioactive fallout. The bombs at Hiroshima and Nagasaki were airburst, and thus produced relatively little fallout. However, if a nuclear bomb is detonated at ground level, tons of radioactive dirt and rubble will be drawn upward into the air. Some of this material will return to the Earth's surface within minutes, landing close to the explosion where people would have been killed or fatally irradiated anyway. However, the remaining particles will circle the globe as deadly radioactive dust before being brought back to Earth by wind, rain, and the force of gravity.

Various commentators have described the effects of a one-megaton bomb exploding 8,500 feet above the Empire State Building in New York: The bomb would destroy virtually every building within a sixty-one-square-mile area and heavily damage those within two hundred square miles. Two miles from ground zero, winds would reach four hundred miles an hour. Jonathan Schell continues,

> Meanwhile, the fireball would be growing until it was more than a mile wide, and rocketing upward to a height of over six miles. For ten seconds, it would broil the city below. Anyone caught in the open within nine miles of ground zero would receive third-degree burns and probably be killed. Closer to the explosion, people would be charred and killed instantly. From Greenwich Village to Central Park, the heat would be great enough to melt metal and glass. Readily inflammable materials such as newspapers and dry leaves would ignite in all five boroughs and west within a radius of about nine and a half miles from ground zero, creating an area of more than two hundred and eighty square miles in which mass fires were likely.

All this from a single, relatively small, one-megaton bomb.

Numbers. They mean nothing. And yet, converted to individual suffering, they are everything. Hiroshima, writes one survivor, was "a mother driven half-mad, looking for her child, calling his name. At last she found him. His head looked like a boiled octopus." Another survivor of the blast remembered "a stark naked man standing in the rain with his eyeball in his palm." . . . "I just could not understand why our surroundings had changed so greatly in one instant," the writer Yoko Ota later recalled. "I thought it might have been something which had nothing to do with the war; the collapse of the Earth, which it was said would take place at the end of the world, and which I had read about as a child."

The survivors of even an isolated nuclear blast would face contaminated water, inadequate sanitation, and a general breakdown of public health leading to outbreaks of cholera, tuberculosis, and typhoid fever. Hospitals would be particularly vulnerable to disruption, since the injuries occasioned by radiation exposure require sophisticated technology; yet in all likelihood, medical facilities near the target city would have been destroyed. Indeed, keeping in mind the worldwide cooperative response to Chernobyl and the extraordinary resources that were necessary to save a relatively small number of people, it's clear that virtually nothing could be done to help the survivors in a setting where, in all likelihood, the doctors would be victims, the hospitals in ruins, telephones, radios, and computers out of order, and casualties in the tens of thousands. A single explosion over Manhattan could doom ten million people, and every major city in the world is equally vulnerable.

None of us like to think about this reality. "Being an ingenious people," Albert Einstein observed, "Americans find it hard to believe there is no foreseeable defense against atomic bombs, but this is a basic fact. No center of population on the earth's surface is secure from surprise destruction in a single attack. There is no defense in science against the weapon which can destroy civilization."

Not everyone accepts Einstein's view. After all, this has been a century of "unacceptable" losses from which civilization has recovered. The Jewish people survived Auschwitz and Dachau, and the state of Israel has been reborn. Europe and the Soviet Union have been reconstructed since World War II. Hiroshima has been rebuilt, which is proof to some that nations can recover from a limited nuclear war. Last year, 25 percent of all Americans believed that in a nuclear war against the Communist world, faith in God would ensure our survival. Of course, political clerics in Libya, Iran, and several other countries fanatically cling to an opposing view.

210

Amidst it all, however, several truths seem certain. The detonation of a single nuclear weapon would be catastrophic. A broader but "limited" nuclear exchange would bring unprecedented suffering to the world. And full-scale nuclear war could lead to the extinction of mankind.

There was a time when those who prophesied the end of the world were derided as paranoid and pictured as placard-carrying psychotics in cartoons. No one of sound mind thought "the end" was near. However, as early as 1956, Dwight D. Eisenhower wrote that in the future both sides in the Cold War would have to "meet at the conference table with the understanding that the era of armaments has ended, and the human race must conform its actions to this truth or die." Eisenhower's view was mirrored in the policies of his successor, John F. Kennedy, whose administration negotiated a treaty with the Soviet Union banning nuclear tests in the atmosphere, outer space, and underwater. Still, stockpiles on both sides have continued to grow to the point where Hiroshima now represents less than one-millionth of the power in the world's nuclear arsenals.

Our planet is finite. Any one of us can get in an airplane and fly around it in a little more than twenty-four hours. Manned spacecraft orbit the globe once every ninety minutes. Our home really isn't all that large, and the release of energy from mass has fundamentally changed the relationship between Earth and man. "In the shadow of this power," observes Schell, "the earth became small and the life of the human species doubtful. . . . The doomsday machine has been assembled. What was once merely a phrase in books is now actuality. . . . These bombs grew out of history, yet they threaten to end history. They were made by men, yet they threaten to annihilate man. They are a pit into which the whole world can fall."

The biological destruction of mankind could occur. We know of no life in the universe other than on Earth, and the extraordinary circumstances that allow for the existence of life are particularly fragile. It's not unreasonable to assume that our planet could be rendered equally in-hospitable as the rest of the cosmos as an environment for living things.

Death would come in waves after an all-out nuclear war. Hundreds of millions of people would die in the initial explosions. Unprecedented famine and plague would follow. True, our species could survive. If only a million men and women (one-fiftieth of 1 percent of the population) were left alive, over time they might re-populate the globe.

However, nuclear weapons do more than kill people. They bring about ecological changes and are capable of devastating the environment to the point where it might no longer support human life at all. Debris from explosions and smoke from fires could rise and blot out the sun, causing a "nuclear winter," with temperatures on the Earth's surface dropping as much as 100 degrees Fahrenheit for up to a year. Further depletion of the ozone layer and extensive radioactive fallout contaminating the globe would also result. Scientists disagree over the amount of megatonnage necessary to trigger these conditions, but the disagreement is over numbers, not principles. Virtually every physicist agrees it is possible to destroy all human life on our planet. Once again, Jonathan Schell puts the matter in perspective:

> It is of the essence of the human condition that we are born, live for a while, and then die. Through mishaps of all kinds, we may also suffer untimely death. In extinction by nuclear arms, everyone in the world would die. . . . From a human point of view, our extinction is an unlimited consequence. It would not only put an end to living generations but foreclose all future generations down to the end of time. It would mark the defeat of all human strivings, all human hopes, all human ideals, past and fu-

ture. . . . Our modest role is not to create but only to preserve ourselves. The alternative is to surrender ourselves to absolute and eternal darkness: a darkness in which no nation, no society, no ideology, no civilization will remain; in which never again will a child be born; in which never again will human beings appear on the earth, and there will be no one to remember that they ever did.

"All of this," Bertrand Russell wrote a quarter-century ago, "is a product of human folly. It is not a decree of fate. It is not something imposed by natural conditions. It is an evil springing from human minds." And the question then is, what should be done about it?

First, we suggest that there be an absolute end to the proliferation of nuclear weapons to countries that don't already have them. The United States isn't adverse to exercising its influence in the international arena, and the Soviet Union is equally "involved." These two governments, acting jointly, should do everything in their power to enforce existing non-proliferation treaties against signatories and non-signatories alike. We simply cannot allow more nations to acquire nuclear weapons.

Second, we propose an end to nuclear weapons tests by all world powers. To date, the United States has conducted approximately seven hundred announced tests. Two hundred and twelve of these took place in the atmosphere before atmospheric testing was banned. Five were underwater. Almost six hundred took place in the continental United States, most of these in Nevada. There's no way to know precisely what effect today's tests will have on future generations. Just as early tests were conducted without regard for the dangers of radioactive fallout, physicists today may be unaware of any number of nuclear perils. Both sides in the Cold War already have enough weapons stockpiled to exterminate mankind several times over, and enough is enough. Indeed, the United States needed only one test to develop the bombs that destroyed Hiroshima and Nagasaki. Both sides should stop testing now.

The world would also be a much safer place to live in if we could find a way to drastically reduce the number of existing weapons in our respective nuclear arsenals. In all likelihood, nuclear warheads will never be abolished. They're part of our heritage. "The physicists have known sin," wrote J. Robert Oppenheimer, director of the Manhattan Project, "and this is a knowledge which they cannot lose." But it's important to remember that nuclear weapons are real. They're not a rhetorical device or imaginary horror we can blithely ignore, and by accident or design, all of us could someday experience Hiroshima. Great disasters seldom arrive in totally unexpected fashion. In the nuclear arena, we've had our warning. Unilateral disarmament would be dangerous and impractical, and it may be that some more advanced defensive system will be desirable in the future to guard against an attack from a Third World nation with a nuclear arsenal more limited than that of the Soviet Union. Still, all citizens should demand meaningful arms reduction. The issue is perhaps the most critical of all time.

All of this, of course, will be impossible to accomplish without a cooperative effort between the United States and the Soviet Union, and this book is in part a plea for that cooperation. In many respects, the Soviets are our adversaries. We have entirely different political and economic systems, and compete with each other in myriad fashions around the world. But we must also remember that the United States and the Soviet Union have never been at war with each other, and few powers of the past millenium can make that claim. Neither side has anything to gain through eternal conflict, and our objectives in achieving nuclear safety are similar if not identical to theirs. The bottom line is, we have to cooperate; there's no alternative. Absent a nuclear holocaust, the Soviets aren't going to disappear and we won't either.

How do nations cooperate to decrease the possibility of nuclear war? They try to understand each other; it's the only way. "To act intelligently in human affairs," Einstein wrote, "is possible only if an attempt is made to understand the thoughts, motives, and apprehensions of one's opponent so fully that one can see the world through his eyes. All well-meaning people should try to contribute as much as possible to improving such mutual understanding." Drawing upon Einstein's view, John F. Kennedy declared, "As Americans, we find Communism profoundly repugnant as a negation of personal freedom and dignity. But no government or social system is so evil that its people must be considered as lacking in virtue. We can still hail the Russian people for their many achievements in science and space, in economic and industrial growth, in culture, in acts of courage."

Perhaps a new era is dawning in the Soviet Union; perhaps not. Regardless, we as Americans have much to gain and little to lose by working to understand the Soviets. In many ways, their philosophy is different from ours. We value our participation in a way of life called democracy. The long-term stability and welfare of our polity are dependent upon the exercise of individual rights. By contrast, the Soviet people traditionally have been willing to subordinate individual rights to the power of the state. Ask the average Soviet citizen, "Do you consider yourself free?" and the answer will be yes. It wouldn't occur to most Soviets to test the boundaries of freedom by writing a dissenting political tract, let alone burning a Soviet flag or turning it into a T-shirt. We in the United States believe that closed borders and censorship as they exist in the Soviet Union are hallmarks of a totalitarian society. But Soviet workers look at our country with equal skepticism and ask, "What about someone who works all his life for General Motors, and then he's fired—is that freedom? Are the people who sleep on the streets of New York free?" It's a very different view of freedom.

We have to understand that most Soviet citizens don't envision themselves as living in a police state. And within the range of human behavior on this planet, in many respects they're quite similar to us. They love their children, their language, their country, nature, music, and art. They want a better way of life for their families and themselves, now and in the future. They want peace, for far more than most Americans, the Soviets in this century have suffered horribly from war. And in a sense, they need peace more than we do because their economy is failing. We can make a full line of consumer products and advanced weapons. They must choose one or the other.

But beyond all this, we speak of cooperation because, as Einstein warned, "a new type of thinking is essential if mankind is to survive." We refuse to accept the notion that war and antagonism will be with us for all time. We believe that someday the United States and the Soviet Union will heal their wounds of hatred in the fashion of other historic adversaries like England and France. This transformation of attitudes may take several generations to accomplish. It will require a commitment from all concerned without regard to changes in political leaders and their whims. But it can be done, and the cooperative effort that followed Chernobyl underscores our optimism.

Americans believe that nothing is impossible. Sometimes we succeed; sometimes we fail; but we always try. Now the time has come for the American people to mobilize populations around the world in an effort to curb the nuclear menace which threatens us all. For too long, too many of us have chosen to do nothing. We've ignored the danger to our survival while the links between complacency and annihilation have grown. We've sought refuge in the hope that an accident won't happen, and allowed a dangerous fatalism to take hold. "We're ordinary citizens," we tell ourselves. "These are issues for scientists

216

and statesmen to resolve." But they're not; they're our issues. Every last person on Earth is involved in the nuclear dilemma. We think "it" will never happen, but if we stay on our present course, someday, by accident or design, it will.

In the late 1960s and early '70s, when a segment of the American population—primarily students of draft age and their families—felt a direct threat from the Vietnam war, they turned the country upside down. The antiwar movement was a "people's movement," as was the push for equal rights, which began in the South and spread nationwide to encompass all ethnic minorities, women, and others whose entitlements were denied. Now, more than ever before in history, a global people's movement is required.

We're heartened by the knowledge that this book will be translated and read in nations around the world. We urge all people to make the same demands of their leaders that we make of ours. Each of us must do what he or she can to halt the nuclear peril; not one of us is exempt from responsibility by virtue of "not understanding" the dangers involved. The evidence is ample. At this juncture in history, anyone who "doesn't know" simply doesn't want to know.

Nuclear weapons present an unprecedented challenge. No previous generation has faced issues as urgent, demanding, and complex in kind. We teeter at the edge of annihilation, but men and women created this risk, and so too can they resolve it. Again, Einstein:

> Science has brought forth this danger, but the real problem is in the minds and hearts of men. We will not change the hearts of other men by mechanism, but by changing our hearts and speaking bravely. . . . For there is no secret and there is no defense; there is no possibility of control except through the aroused understand-

ing and insistence of the peoples of the world. . . .
We need a great chain reaction of awareness and
communication. Proposals should be discussed
in every newspaper, in schools, in churches, in
town meetings, in private conversations, and
neighbor to neighbor. Reading about the bomb
promotes knowledge in the mind, but only talk
between men promotes feelings in the heart. . . .
In this lies our only security and our only hope.
We believe that an informed citizenry will act
for life and not death.

"Our generation," concludes Jeremy Rifkin, "is very
special. It carries the hopes of humanity with it. It is the
lightning rod that connects everyone that has ever lived
with everyone that is yet to be. It is to ourselves that we
must now look for inspiration and guidance. We gain a
measure of immortality by being a nurturing link in the
chain of life. It is by replenishing the rich reservoirs and
deep wells that sustain life, and by carefully stewarding
the earth that bears and gives forth the fruits of life, that
we live on. Our own life force, then, does not die. It is
incorporated into every future birth of every living thing
that results from our nurturing care. By caring for the
creation, we pass on the gift of life, and in the process
our spirit is passed on as well. It becomes part of the
everlasting continuum. When one stops to ponder the
long history of the human sojourn, the millions upon
millions of years in which countless generations have
come and gone, adding their small imprint to the collec-
tive experience of the human race, the mere thought that
our generation will determine whether there will be a
future at all is almost too incredible to comprehend."

Yet our generation has that power. And unless the superpowers of the world act quickly in concert, there will
be a disaster. How many people will die? One hundred
thousand? A million? A billion? All of humanity? Go

back four thousand years to Ancient Egypt and the banks of the Nile. Now move ahead four thousand years to 5988. Does anyone truly think our species will survive till then? How long has man ruled the world? Ten thousand years? Twenty thousand? Will our reign be less than that of previous rulers? Does it matter? Would our extinction have any relevance on a cosmic scale?

The manner in which the nations of the world answer these questions may well reveal whether our presence on Earth is an accident of fate or part of a Divine plan. In the first instance, there would be no one left to appreciate our works; no one to read Shakespeare, listen to Beethoven, or understand our folly. The setting sun would still render the sky ablaze over the Grand Canyon, but there would be no one left to see its splendor.

There are, however, two views of life: the first, that we are all dying; the other, that we are all alive. And if a more peaceful world order evolves out of our nuclear dilemma, then we will have been privileged to live through a great turning point in history—the dawn of a truly Golden Age for mankind.

INDEX

INDEX

INDEX

INDEX

Fahey, John, 112–113
Federation of American Scientists, 204
Fenwal leukopheresis machines, 64, 69–70
Fermi, Enrico, 7–8
Fermi Unit Number 1, 25
fetal liver cells, 54–55, 175
Fetisov, Nikolai, 47–48, 58, 87, 90, 92, 124, 129, 130–131
fission, 4, 22
  military applications of, 8, 10–11, 25
  in nuclear power generation, 14–16
  rapid-fire, 7
  regulating rate of, 15–16
"Five Hundred Years of Masterpieces," 85
forbidden zone, 132
Ford Foundation, 111
Foreign Relations Committee, Senate, U.S., 41
fossil fuels, 14, 182, 194, 201
France, 185
Fulton, Robert, 9
Fyodorov, Svyatoslav, 139–140

Gale, Elan (son), 33, 151–153, 155
Gale, Harvey (father), 105–109
Gale, Mrs. (mother), 105–109
Gale, Robert Peter:
  arrival in Moscow of, 47
  on autopsies, 148–149
  background of, 105–114, 119
  Baranov, personal relationship with, 74, 122, 161–162
  at Chernobyl, 130–135, 152, 170–172
  education of, 107–112
  in Ethiopia, 110–111
  family of, 32–33, 105–108
  Gorbachev's meeting with Hammer and, 92, 93–99
  help offered to Soviet Union by, 34, 36, 42, 173–174
  internship of, 111–112

Gale, Robert Peter (cont.)
  Israeli loyalties of, 154–155
  in Kiev, 123–131, 152–153, 165–167, 170
  leukemia work of, 31–32, 137, 177
  long-term medical evaluation agreement proposed by, 141–142, 143, 147–148, 154, 161
  media and, 45, 64, 68, 84–85, 87–92, 99–100, 102, 142, 147
  T-shirts commissioned by, 143
  at UCLA, 32, 58, 111–114, 115
  Yom Kippur and, 163–164
Gale, Shir (daughter), 33, 45, 155
Gale, Steven (brother), 106
Gale, Tal (daughter), 33, 45, 155
  Soviet trip of, 151–153
Gale, Tamar (wife), 32–33, 49, 64, 74, 77, 82, 84, 88, 101, 102, 114, 116, 145, 151, 152, 154, 155, 166
  American citizenship for, 143–144
  at Yom Kippur service, 163–164
gamma rays, 19, 208
gastrointestinal tract, 34, 175
General Electric, 14
"General Electric College Bowl," 108
generators, 14
genes, regulatory, 21
genetic mutations, 20, 176
geothermal power, 193–194
Germany, Nazi, 7, 9, 10, 25, 128–129
Gershwin, George, 144
Ginsberg, Allen, 106
Goldfarb, David, 167
Goldman, John, 45, 79
Gondar, Ethiopia, 110–111
"Goodwill Games," 147
Gorbachev, Mikhail, 92–99
  on arms reduction, 97–99, 117
  Hammer's letter to, 41–43
  leadership style of, 92–93, 176
  meeting with Gale and Hammer, 94–99
  speech about Chernobyl, 87–88
  on U.S. foreign policy, 96–98, 117
  on U.S. media, 87–88, 95–96

INDEX

INDEX

INDEX

Moscow State Museum of Art, 85, 93
Moscow University, 38
"Mother Russia," 124, 145, 153
mutations, genetic, 20, 176
Muzavieda, Ludmila, 69
myopia, 139

Nagasaki, atomic bombing of, 9–10, 27, 91, 135, 208
effects on victims of, 20
Napoleon III, Emperor of France, 9, 23, 169
National Academy of Sciences, 141, 190
National Cancer Institute, 37
National Gallery, U.S., 116–117, 122
National Guard, U.S., 51
National Hotel, 77–78, 142
National Institute of Health (NIH), 112, 114–115, 141
natural gas, 194
nausea, 20, 55
Nazi Germany, 7, 9, 10, 25, 128–129
neutrons, 5–6
added to uranium, 7, 18
Newton, Isaac, 6
New York Post, 96
New York Times, 107
Nikitin, Ivan, 87, 152, 154
Nixon, Richard M., 38
Nolan, Anthony, Trust, 45
Novik (plant worker), 153–154
Novosibersk, 146–147
nuclear power:
benefits vs. dangers of, 4, 13–14, 16, 21, 22, 182–201
critics of, 182–183
discovery of, 7, 25
efficiency of, 16
government monopoly on, 13
public support for, 183, 216–217
recommendations for, 198–201
as "technological Vietnam," 183
Third World and, 185, 200

nuclear power (*cont.*)
worst-case scenario, 178, 196, 211–212, 218–219
nuclear reactors, 13–18
contingency plans, for accidents, 83–84, 189, 199–200
control room operators at, 186–188
coolants for, 26, 185–186
core of, 15, 17
danger zone around, 199
design of, 24
energy-release rate of, 15–16
first, 14
number of, in U.S., 184
possible meltdown of, 17–18, 24–25, 159, 200
radiation leaks from, 16–17
radioactive waste from, 16, 19, 190–192, 200
regulation of, 13–14
risk of accident in, 185–187, 190, 198
sabotage of, 187–190
staffing of, 198–199
Nuclear Regulatory Commission, 14, 24, 186, 188
nuclear war, 204–219
aftermath of, 211–213
"limited," 211
nuclear weapons, 183, 192–193, 203–219
effects of, 207–210
Einstein on, 7–9
energy yield of, 207
IAEA and, 193
initial ideas about, 7–12
"nuclear winter" from, 212
proliferation of, 11–12, 200, 205–206, 213–214
Schell on, 209, 212–213
testing of, 25, 68, 117, 213–214
see also atomic bomb

Occidental Petroleum, 37–38, 39, 46, 85, 99, 166
oil, 14, 182, 194–195, 201
oil embargo (1973), 193

227

INDEX

INDEX

INDEX